Praise for

Reimagining Payments

"The brilliance of Gitlitz's *Reimaging Payments: The Business Case for Digital Currencies* is to reveal how the coming digital future of value will provide enormous financial efficiencies while disrupting venerable market structures and present, most importantly, the critical challenges and opportunities this change poses for free societies the world over."

—J. CHRISTOPHER GIANCARLO, former chairman, US Commodity Futures Trading Commission

"The leading expert on payments regulation deftly (and downright thrillingly) situates cryptocurrency in historical, practical, and regulatory context—with brilliantly grounded considerations for the future of payments. This book is a powerfully accessible combination of information and inspiration, like the democratization of finance at its best."

—MICHAEL MOSIER, former acting director and digital innovation officer of the US Treasury's Financial Crimes Enforcement Network

"*Reimagining Payments* provides a sweeping view of the payments system. It explains and places current innovations in digital assets squarely in context and highlights the positive real-world impact those technologies can have on the payment system's most vexing problems. This is a great read for payments practitioners, students of economic history, and those interested in digital assets."

—PAUL DWYER, cofounder and chief executive officer, Viamericas Corporation

"Thoughtfully outlined and comprehensive, yet easily readable, *Reimagining Payments* is a primer on digital currencies that everyone from novice readers to industry veterans can benefit from reading."

—JEANINE HIGHTOWER-SELLITTO, chief commercial and strategy officer, EDX Markets

"I wholeheartedly recommend *Reimagining Payments* to anyone intrigued by the world of digital transactions but who might find the subject too daunting to navigate. Gitlitz's expertise shines through, making this book an indispensable, accessible guide for understanding the past, present, and future of payments."

—DAVID PUTH, former CEO of CLS Group and Centre Consortium and former J. P. Morgan executive

"With a deep understanding of technology policy, cryptocurrency, and fintech, *Reimagining Payments* is an essential read. The book not only debunks common misconceptions about cryptocurrencies but also illuminates their growing commercial significance. Gitlitz's work is a rich source of strategic guidance and practical insights, making it indispensable for professionals and enthusiasts keen on understanding the future of digital payments."

—PAUL BRIGNER, head of policy and strategic advocacy at Electric Coin Co. and adjunct fintech professor at Georgetown University McDonough School of Business

"I thoroughly enjoyed this easy-to-read passage through disruptions in payments and comparisons to the dot-com era."

—BRUCE SILCOFF, cofounder, Shyft

"Not your typical finance read, this is an homage to money's historic arc, wrapped in practical insights on the future of payments. With charm and wit, Michelle Gitlitz transforms technological complexity into a compelling narrative that sparks creativity and equips you with everything you need to confidently navigate the evolving landscape of digital payments."

—JESS CHENG, member at Wilson Sonsini Goodrich & Rosati

"Michelle has been an industry leader in the field of digital currencies and payments for many years. This book will paint a picture for you of how the world of money and payments has been disrupted by emerging technologies and, most importantly, what the future of payments holds."

—JASON BRETT, Digital Assets Senior Contributor, Forbes.com

Reimagining Payments

Reimagining Payments

The Business Case for Digital Currencies

Michelle Ann Gitlitz

RACKET

Gilbert, AZ

Reimagining Payments: The Business Case for Digital Currencies

For information about this title or to order other books and/or electronic media, contact the publisher:

Racket Publishing | www.racketpublishing.com

Ebook ISBN: 979-8-9878649-0-6
Paperback ISBN: 979-8-9878649-1-3
Hardcover ISBN: 979-8-9878649-2-0

Printed in the United States of America

Interior design: Jessica Angerstein

For Emily and Amelia

Never give up.

"We tend to overestimate the effect of
a technology in the short run and
underestimate the effect in the long run."

—AMARA'S LAW

"Innovation distinguishes between
a leader and a follower."

—STEVE JOBS

Contents

List of Figures and Tables

Introduction

"Time is a flat circle."

—FRIEDRICH NIETZSCHE

Those of us of a certain "vintage" remember that, in the late 1990s, dot-com start-ups were all the rage. Companies with dubious or nonexistent business models raised tons of cash from excited investors. Some went public to stratospheric valuations that, in hindsight, made little sense. New-economy zealots touted eyeballs, time-on-site, clicks, and other newfangled metrics to justify inflated valuations. The internet was going to change everything, and fear of missing out (FOMO) drove a frenzy of questionable investments and business decisions.

Here's a case in point. Launched in 1995 as AudioNet, Broadcast.com was one of the first streaming services. It was premised on allowing out-of-town sports fans the ability to listen to their favorite teams' games over the internet. The site was popular and convenient. It stood to reason, then, that these attributes would *eventually* generate big bucks. Mark Cuban, an Indiana University Hoosier who missed his alma mater's college basketball games and saw the promise of webcasting, invested in the company. He took over its management and led

Broadcast.com through an IPO in 1998. At the time, it had less than $7 million in revenues, $28 million in equity, and an accumulated deficit of nearly $10 million. Achieving profitability wasn't in the company's foreseeable future. Yet, on April 1, 1999, then Yahoo! CEO Timothy Koogle acquired Broadcast.com for a whopping $5.7 billion in stock.[1] Only two years later, Yahoo! shut down much of its broadcast services, and Broadcast.com was discontinued. Yahoo!'s purchase of Broadcast.com has since been called one of the worst internet acquisitions of all time.[2]

The growth of the internet created a buzz among investors, who were quick to finance these companies, notwithstanding that some lacked a business plan, product, or track record of profits. The dot-com crash demonstrated that tried-and-true business fundamentals still mattered—even if the market seemed to be ignoring them. Eventually, the bubble had to pop, and it did.

Cracks Surface

On February 11, 2000, online retailer Pets.com began trading on the Nasdaq at $11 per share. Despite being nowhere near profitability, the excitement of the era pushed the stock to $14.

Pricey Super Bowl commercials* and a cute mascot wouldn't translate to profits and a positive cash flow. This emperor had no clothes, and its shares quickly plunged below a dollar. The company folded in November 2000, firing roughly 300 employees in the process.[3] Other dominoes quickly started falling. Companies such as Boo.com, Global Crossing, eToys, and WorldCom collapsed.

By the start of the new century, the macro sentiment had shifted, and investors were finally starting to come to their

* Watch one of them at https://tinyurl.com/repay-pets.

senses. Correction time had arrived in earnest. The bursting of the bubble caused market panic through massive selloffs of dot-com company stocks, plunging their values further. On March 12, 2000, the Nasdaq fell more than 9 percent. It dropped an additional 10 percent one month later. By 2002, estimated investor losses were around $5 trillion.

Execs at brick-and-mortar retailers such as Blockbuster, Tower Records, and Borders Books breathed a collective sigh of relief.

A Postmortem

Were the dot-com naysayers right?

Sort of. The answer is nuanced.

Share prices of internet companies increased much faster and higher than their non-internet peers due mostly to speculation caused by the promise of the new internet technology. More than two decades after the dot-com bust, however, Amazon, eBay, Netflix, and Google remain some of the world's most valuable, successful, and influential corporations. Why? Their intrinsic worth supports their valuations.

As it turned out, people love buying things online—and not just books. In 1999, Zappos, which was formerly shoesite.com, pioneered home delivery of shoes—with free returns to boot. Despite its recent financial struggles, Carvana sold more than 400,000 used cars online in 2021 alone.[4] Think about it: try to name a *legal* product that you'd like to buy online but can't.

I bet you're hard-pressed to find one.

Brass tacks: reports of the demise of ecommerce were exaggerated. Companies that didn't adapt have gone *kaput*. Blockbuster, Tower Records, and Borders have all gone the way of the flightless dodo. Walmart, Target, and other big-box retailers had to up their online presence to survive.

It's not the first time that excitement about a new technology got ahead of economic reality. Despite the dot-com bust, many viable use cases—including online communications, shopping, and digital entertainment—survived and thrive to this day because of amazing new technology and fundamental shifts in consumer behavior.

Lessons, Parallels, and Cryptocurrency

The Great Recession and subprime mortgage crisis of 2008 spawned the rise of Bitcoin and other cryptocurrencies, as I discuss in Chapter 4. In July 2023, global cryptocurrency market capitalization was $1.19 trillion.[5] But 2022 was a pretty brutal year for the industry. Cryptocurrencies seemed to be all the rage during the early part of the year, with celebrities endorsing various companies on the world's largest advertising stage—the Super Bowl.

However, like just about everything else in finance, cryptocurrency prices plummeted when the Federal Reserve started to raise interest rates to fight high inflation as 2022 progressed. A string of industry failures started in May 2022, including the implosion of the Terra and LUNA cryptocurrencies, the trading platform Voyager, the crypto hedge fund Three Arrows Capital, and the lender BlockFi. And we can't forget the implosion of FTX in November 2022. The inevitable crypto winter had arrived.[6] The price of Bitcoin—a decent proxy for the overall cryptocurrency market—plummeted.

Yet, as I write these words in December 2023, the price of Bitcoin has recovered nicely from its nadir. Explanations for industry resurgence run the gamut. The industry has gone through other steep downturns since Bitcoin was introduced in 2009. We saw a prior crypto winter in 2018 after a flurry of initial coin offerings, in which start-ups in the ecosystem raised tons of money based on shaky valuations.

Fundamentally, blockchain technology, which enables cryptocurrencies, has the potential to revolutionize numerous industries. One use case—and the subject of this book—is payments. Blockchain technology and cryptocurrencies solve a plethora of thorny payment problems, including cost, security, privacy, and chargebacks that have long plagued the sector.

Cryptocurrencies are here to stay, and it's high time to reimagine payments.

Michelle Ann Gitlitz
December 1, 2023

Part I:

The Rudiments of Finance, Payments, and Banking

Chapter 1

You Think You Know Money

"A nickel ain't worth a dime anymore."

—YOGI BERRA

Let's light this candle with two big questions: what is money, and what does it do?

Economists answer those questions in several ways.

The Meaning of Money

First, money serves as a store of value or wealth. In other words, your dollars generally maintain their value and purchasing power over time. This is perhaps money's most essential function, and recent actions by the Federal Reserve only underscore this reality. (Since early 2022, the Fed has aggressively raised interest rates to combat the pernicious effects of inflation to avoid a repeat of the issue-plagued 1970s. Most industrialized countries have followed the Fed's lead.)

In this context, contrast money with depreciating assets. A new car loses anywhere between 9 and 11 percent of its value the moment its owner drives it off the lot.[1] Imagine if your paycheck did the same. Widespread chaos would ensue.

Then imagine a local restaurant that couldn't run a simple, accurate profit-and-loss (P&L) statement because its owner couldn't easily determine the value of its assets and liabilities.

In this way, money serves as a unit of account and a common measure of value across the economy. Thanks to money, the restaurateur can easily determine her debits and credits—her accounts payable and accounts receivable.

On an individual level, think of money as a key mechanism for a human being to make transactional decisions and exercise agency. Money helps us measure value in all sorts of economic transactions. It allows us to determine if we should splurge an extra $700 on that first-class upgrade for a cross-country flight or sit in coach.

Now think broader.

In effect, no contemporary economic system can exist without money, especially capitalism. No, it's not perfect. (To paraphrase Churchill's iconic line about democracy, it's the worst system—save for all the others.) In the words of J. H. Cullum Clark, director of the Bush Institute-SMU Economic Growth Initiative:

> If it's working right, the free-market system produces goods and services better than any alternative. It creates powerful incentives to innovate, and generally ensures people's earnings reflect the value they deliver to others through work.[2]

Finally, money enables payments. As you probably guessed from the title of this book, I'll focus in the coming pages on this particular function of money.

Money is a means of exchange—historically, legal tender widely accepted between and among parties as a method of payment. For example, if you decide to see *Taylor Swift: The Eras Tour* in Manhattan or Phoenix, the AMC or Regal theater will accept US dollars as payment. (Interestingly, both chains also accept certain digital currencies, even though they're not legal tender. Chapter 3 will return to this topic.)

Money Matters

If money is supposed to accomplish essential economic, financial, societal, and legal objectives, the next natural question becomes, how?

Sovereign governments typically create and issue money through central banks. This is a critical point: there's no official world currency, although the US dollar has served this de facto function for decades.

Our multinational world means that money is neither universal nor available everywhere to everyone. Whether you're in Alabama or Wyoming, US dollars work just fine. Head to Sweden, though, and you'll need to find some krona if you're intent on paying for that Big Mac with cash. Ditto for the peso in Mexico and the yen in Japan.

Types of Money

All money takes the form of either tokens or accounts. To summarize the key difference between the two, consider the words of Charles Kahn, a research fellow with the Federal Reserve Bank of St. Louis:

> When you pay with an account, the crucial question for the recipient is your identity: "Are you really the account holder?" When you pay with a token, your identity is irrelevant; instead, the crucial question for the recipient is: "Is this object I'm receiving real or counterfeit?"[3]

Let's unpack these two types of money.

TOKENS

It's instructive to think of hard cash and coins as fungible tokens—that is, nonunique items that represent value. Economists liken fungibility to interchangeability. One bushel of wheat is the same as any other; ditto for a barrel of oil. Provenance is irrelevant. The same holds true for a generic quarter or five-dollar bill. (Coin collectors get a pass here.)

Say that you're putting that quarter into a *Ms. Pac Man* machine at an arcade, as I loved doing as a kid. That coin is a local instrument that allows you to complete a physical transaction—in this case, playing a video game. The arcade owner doesn't need to verify your identity to complete the transaction. The inanimate machine doesn't care who you are; it only needs to verify that the coin is real and not counterfeited. What's more, the money is available to you 24/7/365, although you can only use it during the arcade's hours of operation.

ACCOUNT-BASED MONEY

Contrast anonymous, verification-free tokens with account-based money. The latter relies on verified identities. They represent account liabilities on the balance sheets of banks and other financial institutions. The use of account-based money for payments requires confirmation of each of the following:

- The identities of all the parties in the transaction.
- The adequacy of funds by the payer or transferor.
- The delivery of funds to the recipient.

A brief example will demonstrate how, if you remove any of these elements, the payment fails.

Let's say that I typically pay my dry cleaner with cash, but today my wallet is empty. I write her a check.

1. She visits her local Chase branch and attempts to deposit the check.
2. If I've overdrawn my account, the check will bounce. (My dry cleaner will likely give me an earful, and I doubt that she'll ever let me pay by check again.)
3. If I have overdraft protection, Chase will cover the shortfall in funds and assess me a fee for the failed transaction.

As you may have guessed, checks are a dying breed. They currently represent only about 7 percent of all US payments.[4] Checks remain instructive, however, in demonstrating the value, shortfalls, and complexities of account-based money.

No bank today processes transactions manually. On the contrary, banks rely upon extensive systems to track all account-based transactions.[5] Despite the technological infrastructure, account-based money suffers from a number of limitations. Much like the arcade maintains regular business hours, account-based money is generally available only during bank operating hours and requires verification for use. To this extent, it's exclusionary, expensive, and slow.

Note that Chapter 2 delves deeper into the history of payments.

An Insanely Brief and Incomplete History of the US Dollar

Money is an abstract concept. Paul Vigna, a reporter who pioneered coverage of Bitcoin and cryptocurrencies at the *Wall Street Journal* and the coauthor of *The Age of Cryptocurrency*, explains why better than I ever could. As he wrote on his Substack in May 2023:

> Take $20 out of your wallet. Get a piece of paper and write "$20" on it. Hold them up to each other. What are they? Quite literally, they are two pieces of paper with some ink on them. Intrinsically they are identical. Now, go to McDonald's and order a Happy Meal. When it comes, hand the cashier the $20 bill. The next day, go back to McDonald's, order another Happy Meal, and hand the cashier the piece of paper with "$20" written on it.
>
> Think you'll get that second Happy Meal?
>
> This isn't just me saying it. Aristotle worked all this out 2,500 years ago. Money wasn't a natural phenomenon, he said, but a shared illusion. Its power was ultimately derived from the law, nothing else. The whole idea of money as a

> scarce commodity, Aristotle argued, is just part of making it feel real. That's why gold works well as a representation of money, he said. It's scarce, and it's got value. But gold isn't money, any more than paper is. Go to McDonald's with a little gold rock and order a Happy Meal. When it comes, pull out a Buck knife and shave a little sliver of gold off the rock. Think you'll get that Happy Meal?[6]

Let's fast-forward a few millennia ...

With a few temporary exceptions around wars and embargoes, the US was on a gold standard from 1879 to 1933. Citizens could redeem their notes for the dense, valuable metal. But on June 5, 1933, Congress passed a joint resolution nullifying the right of creditors to demand payment in gold. President Franklin Delano Roosevelt concurred and, with the stroke of a pen, took the US off the gold standard. Bank failures and the Great Depression had caused widespread panic. The American public had begun hoarding gold. Maintaining the policy became untenable. US citizens could no longer redeem Federal Reserve notes for gold, silver, or any other commodity—and they still can't.[7] At this point, only foreign governments can redeem dollars for gold. (More on this point later in the chapter.)

The decision made many queasy. After all, without the support of an objectively valuable material, how would people assign value to the dollar? As the following sidebar illustrates, it's a question that has intrigued countless great minds.

On Diamonds, Water, and Value

Money's newly "squishy" value may make you a smidge uncomfortable, but the pursuit of "objective value" is a fool's errand. Don't believe me, though. Philosophers Plato and Nicolaus Copernicus and economists John Locke, John Law, and Adam Smith all wrote about the paradox of value—aka, *the diamond–water paradox.*

In *The Wealth of Nations*, Smith compared the high value of a diamond, which is unessential to human life, to the low value of water, without which humans would die. He determined value-in-use was irrationally separated from value-in-exchange.[8]

We can't live without water, but it's effectively free. By contrast, diamonds command high prices but, for the purposes of survival, simply don't matter. (Pro tip: never make that argument to your fiancée.)

The answer to the paradox ultimately comes down to scarcity: the wide availability of water (I'm putting climate change aside here) relative to demand means that the price of water is low or negligible. Diamonds, on the other hand, are relatively rare and expensive to produce, making supply limited. That explains the price differential between water and diamonds even though water is essential to human survival.

WWII AND THE SEARCH FOR STABILITY

Skip ahead to September 2, 1945. World War II had come to an end, and the United States had emerged relatively unscathed. Other than Pearl Harbor, no attacks had taken place on American soil.

The post-war powers were resolute: they were intent on avoiding the mistakes that gave rise to Adolf Hitler in Germany and Benito Mussolini in Italy. World leaders were acutely aware that the First World War had led to an era of intense financial volatility and civilian unrest. The effects of the Allies' punitive policies reverberated for years after its end. Case in point: in 1914, the exchange rate of the German mark to the American dollar was about 4.2 to 1. Nine years later, it was 4.2 *trillion* to 1.[9]

This brings us to the development of the Bretton Woods system in 1944. The Bretton Woods Agreement and system created a collective international currency exchange regime that lasted from the mid-1940s to the early 1970s. Writing for *Barron's*, Randall Forsyth nicely encapsulates the overarching objective of the monetary system for the new world order:

> The Bretton Woods system, in effect back then, reflected America's economic pre-eminence after World War II. Currency exchange rates were fixed, relative to the dollar, which, in turn, was exchangeable for gold at a fixed $35 an ounce. The idea was to avoid the currency instability and competitive devaluations of the 1930s, but with greater flexibility than allowed under the classical gold standard, which most economists agreed had helped trigger and spread the Great Depression.

Bretton Woods largely worked. It "was successful in bringing about exemplary and stable economic performance in the 1950s and 1960s."[10] The Bretton Woods Agreement created two institutions: the International Monetary Fund and the World Bank. Both institutions have withstood the test of time, globally serving as important pillars for international capital financing and trade activities.[11]

NIXON ENDS BRETTON WOODS

Fast-forward to August 15, 1971. Against a backdrop of increasing inflation and the Vietnam War, Nixon wasn't winning any popularity contests. On that day, the thirty-seventh president took a number of drastic economic measures—termed the *Nixon shock*—which were the catalyst for the stagflation of the 1970s as the US dollar dropped in relative value. He addressed the American people in a speech that's worth quoting at length:

> The third indispensable element in building the new prosperity is closely related to creating new jobs and halting inflation. We must protect the position of the American dollar as a pillar of monetary stability around the world. In the past seven years, there has been an average of one international monetary crisis every year ... I have directed Secretary Connally to suspend temporarily the convertibility of the dollar into gold or other reserve assets, except in amounts and conditions determined to be in the interest of monetary stability and in the best interests of the United States. Now, what is this action—which is very technical—what does it mean for you? Let me lay to rest the bugaboo of what is called devaluation. If you want to buy a foreign car or take a trip abroad, market conditions may cause your dollar to buy slightly less. But if you are among the overwhelming majority of Americans who buy American-made products in America, your dollar will be worth

> just as much tomorrow as it is today. The effect of this action, in other words, will be to stabilize the dollar.*

Nixon announced that the US had unilaterally ended Bretton Woods. He directed Treasury Secretary John Connally to suspend the ability of foreign governments to convert the US dollar into gold or other reserve assets, except in certain limited circumstances.

To call Nixon's move *bold* is the acme of understatement on several levels. First, colloquially, people equated gold and money with value for centuries. Case in point: consider the expression "worth its weight in gold." The metaphor first appeared in English in the 1300s, but its etymology dates back to Roman times.[12]

Second, the end of Bretton Woods led to the instability of floating currencies. The US dollar sank by a third during the 1970s, and over the past forty years, it has had several periods of severe volatility.

Third, Nixon transformed the global economy in one fell swoop and ushered in the era of *fiat currency*.†

FIAT CURRENCIES

By way of background, the Latin word *fiat* means "let it be done." When used as a noun in English, it signifies "an authoritative or arbitrary order or decree." Generally speaking, fiat currency is government-issued money that isn't backed by a physical commodity, such as gold or silver. Instead, it's backed by the government that issued it. It's only as valuable as the

* Watch it yourself at https://tinyurl.com/mgnixon.

† For more on this decision and its ramifications, check out Jeffrey Garten's excellent book *Three Days at Camp David*.

creditworthiness of the issuing government. For example, the US dollar is backed by the "full faith and credit of the US government." Don't take my word for it, though. From the US Department of Treasury's website:

> Federal Reserve notes are not redeemable in gold, silver, or any other commodity, and receive no backing by anything. Redeemable notes into gold ended in 1933 and silver in 1968. The notes have no value for themselves, but for what they will buy. In another sense, because they are legal tender, Federal Reserve notes are "backed" by all the goods and services in the economy.[13]

Put differently, there's a social, reputational, and trust-based value to all fiat currencies. The US dollar has been a fiat currency for more than half a century; it's valuable because the US government says so and the world believes it. As of this writing, there are currently 180 fiat currencies in the world.[14]

Because fiat currencies aren't pegged to the value of tangible assets, governments can print as much of them as they want. This reality makes the US dollar and its ilk more susceptible to inflation and deflation.

Meet Bernard von NotHaus

Now that we know more about what money does, it's time to meet Bernard von NotHaus. He's an ex-hippie, the founder and "high priest" of the Free Marijuana Church of Honolulu, and an avid gold enthusiast.

von NotHaus was none too thrilled with Richard Nixon in 1971, but not for reasons that you might expect. The decision to exit Bretton Woods didn't sit well with him.

In September 1974, von NotHaus founded the National Organization for the Repeal of the Federal Reserve Act and the Internal Revenue Code. NORFED first issued currency called the Liberty Dollar on October 1, 1998.[15] Unlike US dollars, Liberty Dollars were backed by gold and silver.

The idea caught on. Over the years, NORFED created and distributed more than 60 million Liberty Dollars.[16] An impressive feat, to be sure, but was it legal?

The US government didn't think so. On December 18, 2006, the FBI arrested von NotHaus. He was charged with making, possessing, and selling Liberty Dollars for use as legal tender to compete with the US dollar. He was also charged with conspiracy against the United States.

Four years later, the case of *United States v. Bernard von NotHaus* went to trial, prosecuted by Anne M. Tompkins, then US Attorney for the Western District of North Carolina. The government's case rested in part on Article I, Section 8, Clause 5 of the Constitution, which grants the government the exclusive right "to coin Money, regulate the Value thereof, and of foreign Coin, and fix the Standard of Weights and Measures."

The government argued that von NotHaus was an illegal counterfeiter who committed a unique form of domestic terrorism by undermining the legitimate currency of the United States.

As expected, von NotHaus disagreed. His lawyers argued that Liberty Dollars were privately issued collectibles, not legal tender. They also reminded the jury of the rich history of private coinage in America—one that dated back to the mid-eighteenth century.[17] von NotHaus likened the relationship of the Liberty

Dollar and the US dollar to FedEx and the US Postal Service. Didn't FedEx offer a *complementary*, private-sector alternative to the public-sector mail system?

What was wrong with creating a private currency for people to voluntarily use to exchange value outside the government's purview? What if free citizens wanted to use it to barter? The US Constitution doesn't prohibit private issuance of money.

Ithaca Is Gorges!*

In 1991, a local businessman named Paul Glover introduced an alternative currency called the Ithaca HOUR. An Ithaca HOUR was valued at US$10 and was recommended as roughly equivalent to payment for one hour's work.

The Ithaca HOUR was successful, with several million dollars' worth of HOURs traded between residents and more than 500 local businesses in the twenty years following its inception. Today the Ithaca HOUR is no longer widely used, but efforts are underway in Ithaca to revive its use.[18]

Unlike von NotHaus, Paul Glover was never accused by the federal government of any wrongdoing.

But back to von NotHaus. Despite the protestations of von NotHaus's legal team, a jury in Statesville, North Carolina, in March 2011 found the sixty-seven-year-old guilty of:

* Cornellians like me will appreciate this reference. For those who are unfamiliar with the phrase, it's a play on words emphasizing the natural beauty of Ithaca's 150 waterfalls carved out by glaciers that formed gorges millions of years ago.

> making coins resembling and similar to United States coins; of issuing, passing, selling, and possessing Liberty Dollar coins; of issuing and passing Liberty Dollar coins intended for use as current money; and of conspiracy against the United States.[19]

Jurors needed all of ninety minutes to deliberate. Needless to say, the government agreed with their verdict. In the prosecutor's words, "Attempts to undermine the legitimate currency of this country are simply a unique form of domestic terrorism." von NotHaus ultimately received a term of probation with six months of home detention.[20]

Now that we've covered the basics of money, it's time to explore one of its most basic functions: payments. That worthy objective will be our starting point for the next chapter.

Chapter Summary

- One of the most essential functions of money is to serve as a medium of exchange.
- Contemporary money falls into two buckets: token-based and account-based. There's a need for both.
- The US dollar is a fiat currency backed by the full faith and credit of the US government, not gold or silver.
- The US Constitution doesn't prohibit private issuance of money, but counterfeiting US dollars is a crime.

Chapter 2

Payments Make the World Go Round

"One pays for everything. The trick is not to pay too much of anything for anything."

—JOHN STEINBECK

On September 12, 1997, a three-year-old ecommerce company filed a patent for a "method and system for placing a purchase order via a communications network."[1] Two years later, the United States Patent and Trademark Office granted the application and issued patent US5960411A.[2] Hello, Amazon.

In a way, patent US5960411A was unremarkable. After all, the USPTO approved 169,085 patents that year.[3] But in another, more accurate way, Amazon's 1-Click* changed everything and catapulted the company from an online bookstore to a massive online marketplace. The technology vastly reduced the friction that consumers had experienced when attempting to pay for

* As you'd expect, Bezos and company also intelligently trademarked the term.

items online. It allowed consumers to enter their billing, shipping, and payment information just once. After that, they could simply click a button to buy something.

USPTO typically issues patents for twenty years, but few prominent companies were willing to wait until 1-Click entered the public domain on September 12, 2017. The tech was too valuable; the need to minimize payment-related friction was too great. Case in point: Apple licensed Amazon's 1-Click technology three years later, in September 2000.[4]

Since its inception, Amazon has never taken its foot off the pedal. (As ex-CEO and company cofounder Jeff Bezos once famously quipped, "Your margin is my opportunity.") In the years since patenting 1-Click, Amazon has continued to push the envelope by disrupting the payments sector. Because this is a book on payments, I'll ignore the massive contributions the company has made to cloud computing, smart homes, publishing, and logistics.

Amazon's subsequent payment-related innovations include Zero-Click Ordering for such things as toothpaste, beef jerky, and soap. The technology automatically determines what you want to buy and ships the product based on an ecommerce algorithm. If you like the jerky, you keep it and your stored credit card is charged. If you have enough jerky, you send it back in the box free of charge. The advent of Alexa allowed for voice-based ordering. Amazon Go and Amazon Go Grocery, the company's burgeoning lines of checkout-free stores, permit Just Walk Out payment technology. That technology marvel "lets shoppers enter a store, grab what they want, and get going."[5] By the way, retailers looking to improve the customer experience can also license Just Walk Out.[6] Each of Amazon's payment inventions has reduced the friction that customers experience when purchasing the company's increasing array of goods and services, thereby exponentially increasing its market share and profits.

Of course, paying for items hasn't always been effortless, and not every merchant accepts a wide array of payment choices. In fact, payments can still be problematic. Plenty of friction exists, even for members of industrialized societies. In the third world, payments still represent a struggle for billions of people.

This chapter explores the history of payments and the source of that friction. We'll see how payments have evolved—often thanks to nascent technologies. Before we do, however, we need to take a step back and cover some basics.

Disclaimer

We cover payments and payment systems here, but only at relatively high levels. We don't need to get under the hood too much for the purposes of this book. The mechanics of payment systems are, well, complicated. Amazon currently lists more than a thousand book titles on payment systems, some of which are 400-page tomes.

Still, this book is called *Reimagining Payments* for a reason. Only fools would reimagine a critical process they don't fully understand. As such, a fair level of detail is necessary for our purposes. This information will serve as a foundation for Part II of this book.

Don't worry. We won't spend too much time in the weeds. I've included subheadings so you can skim this chapter if you want.

The Importance of Payments

We know from Chapter 1 that money serves a number of critical purposes, one of which is to facilitate payments. But what's the definition of a payment? The Merriam-Webster dictionary defines it like this:

1. The act of paying.
2. Something that is paid: pay.
3. Requital.[7]

Black's Law Dictionary defines payment as follows: "The performance of a duty, promise, or obligation, or discharge of a debtor liability, by the delivery of money or other value. Also, the money or other thing so delivered."[8]

Money isn't the only form of legally recognized payment. Debtors can pay with *anything* the creditor willingly accepts to settle the debt. Here's an example.

Say that I go out to dinner and forget to take my wallet and phone with me. When the waiter brings me the check, I realize that I lack a way to pay. If I can convince the creditor (in this case, the restaurant owner) to accept my purse in lieu of cash, legally speaking we're copacetic.

Few of us likely pay close attention to payments. Make no mistake, though, payments are integral to a smoothly running society. Without a means of quickly settling debts, any modern economy would grind to a halt. The following sidebar illustrates how the inability to pay can disrupt life.

A World Without Payments

Nora works as a vice president at a movie studio. On Sunday she intends to fly from her home in Los Angeles to New York for a meeting taking place on Monday morning. Upon arriving at John F. Kennedy Airport, she'll grab an Uber, check into her hotel, and crash. In the morning, Nora will order breakfast, check out, head to her meeting, and then fly home.

When Nora lands in Queens, she discovers that she's the victim of identity theft and that her bank and credit card companies froze her accounts. In short, she can't access her funds. Without a verified payment method, her trip just became exponentially more difficult. She'll have to rely on the cash in her wallet—if it's even accepted—and eventually, she'll run out.

Bottom line: Nora won't be able to resume her normal life until she resolves her payment issues.

Now multiply Nora's situation by 335 million—the approximate population of the United States. To continue with this example, hotels, Uber drivers, and restaurants that can't make and accept payments will quickly shut down.

Fortunately, most Americans don't have to deal with this apocalyptic scenario. However, a decent number do. In 2021, the Federal Deposit Insurance Corporation and the US Census Bureau reported that 5.9 million US households—4.5 percent of the population—are unbanked and *must* rely upon cash.* In other words, in these households, no one can access a checking or savings account at a bank or credit union. (We'll revisit these folks in Chapter 5.)

Generally speaking, smooth payments facilitate commerce. All things being equal, when they work effectively, economies run more efficiently.

The converse is also true: payments that are inefficient because they're slow, costly, and fraught with fraud inhibit commerce and stifle economic progress.

When Cash Was King

Let's briefly travel back in time to January 1, 1950. Harry S. Truman ran the country, Vaughn Monroe's "Riders in the Sky" topped the charts, and the New York Yankees had recently defeated the Brooklyn Dodgers in the World Series.

Pretend for a moment that cash represents the only form of payment. For the sake of simplicity, forget checks, precious metals, and IOUs; credit and debit cards don't exist yet, never mind Venmo, PayPal, and the other payment methods I'll cover later in this chapter.

In this cash-only world, you know the tellers at your local bank. When you need to make a withdrawal, you present valid identification to a human being. What about automated teller machines? (The first ATM didn't arrive until September 2, 1969.[9]) Odds are that you own a safe as well; there's no sense in making a special trip to the bank because you need $10.

* The census typically misses noncitizens, so the number of unbanked residents in the US is likely higher.

A week later, you visit relatives in Toronto. Soon after landing, you find a local bank to procure some Canadian dollars. You plan on attending tonight's hockey game (the Maple Leafs are playing your New York Rangers), and you need to buy a ticket. You know that you can't pay for the ticket with US dollars.

You arrive at the arena at 6 p.m. and have time to spare before the hockey game begins at 7:30. You hand the ticket agent Canadian dollars, and she promptly hands you a ticket. The transaction is *synchronous*. In this case, by definition, there's no risk of the transaction being incomplete or invalid. (You aren't attempting to pay for the ticket using counterfeit bills, and the agent isn't selling you a fake ticket.)

Moreover, the example demonstrates something even more important: today's economy is global, but payment conventions are local. Whether you're in Alabama or Wyoming, merchants will accept your US dollars. But the same doesn't apply to your typical cross-border or overseas merchant.

This simple, cash-only world is appealing on a number of levels. Cash continues to provide universal access to payments. Recipients can immediately access the proceeds from cash-based sales. What's more, using cash to settle a debt doesn't depend on a third party extending credit. There's no additional charge or transaction fee for usage. Finally, cash transactions are anonymous. No one needs to know that you bought that pack of cigarettes or visited a marijuana dispensary—except, of course, the cashier.

Despite the manifold benefits of cash, its halcyon days are in the rearview mirror.

CASH IS NO LONGER KING IN PAYMENTS

Money has always evolved. Ditto for payments. How we pay for goods and services is constantly changing. Specifically, cash is no longer king.

As Malcom Harris writes for *New York* magazine,

> During the first decades of the 21st century, cash has gone from the primary American form of payment to third place. Debit cards jogged past in 2018, and credit cards followed in the first pandemic year, 2020.[10]

In October 2022, Pew Research Center released some interesting survey results on our relationship with money. In less than a decade, the share of Americans going cashless in a typical week has increased by double digits. Figure 2.1 shows just how quickly Americans have eschewed cash.

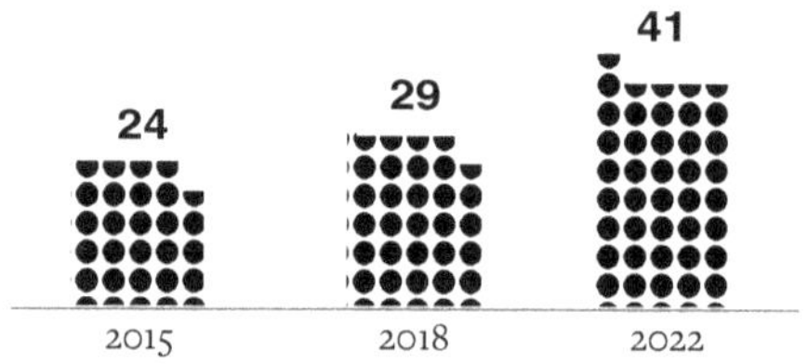

Figure 2.1: The Rise of the Cashless Consumer (Percentage of Americans Who Say That None of Their Typical Weekly Purchases Involves Cash)
Source: Pew Research Center

Pew is hardly alone in reporting on the diminishing importance of cash. In October 2022, the management consulting firm McKinsey released its latest research on the size of the global payments industry.[11] The numbers are astounding:

- COVID-19 caused a dramatic decrease in the use of cash; it plummeted by 15 percent in 2020.
- Even as physical stores reopened in 2021, cash usage increased by only 1 percent. The majority of payers continued to use electronic payments.

In a word, *wow*. Like telemedicine, online education, and ecommerce, COVID-19 accelerated the move away from cash to electronic forms of payment.

It's not hard to understand why cash is no longer king. Count among its many drawbacks:

- Vulnerability to theft.
- Lack of tracking. (It's not as if "cash" sends you a monthly statement detailing your purchases.)
- Incompatibility with modern commerce. Try paying your Uber driver or Amazon delivery driver with greenbacks.
- Lack of practicality. Similar to the first drawback, do you really want to bring $2,000 in bills to the Apple Store to buy that sleek MacBook Pro?
- No ability to earn rewards points or frequent flyer miles.
- Paying with cash precludes people from establishing credit for future big-ticket purchases, such as houses and cars.[12]

THE BENEFITS OF ELECTRONIC PAYMENTS

For consumers, the benefits of electronic payments are too big to ignore. As it turns out, we're not the only ones who benefit from the increased usage of credit and debit cards. The trend is music to the ears of the financial institutions that profit from card-associated fees.

That same McKinsey report noted that revenue from global payments in 2021 was $2.1 trillion, up 11 percent from 2020. By 2026, McKinsey projects that number may hit—wait for it—a mind-boggling $3 trillion.[13]

Banks and other financial institutions earn money in all sorts of ways, but the most relevant for our purposes is by facilitating payments. We've now arrived at the rails—arguably the most essential part of the modern payments universe.

Payment Systems and Intermediaries: Essential Infrastructure

Because we've moved away from cash-based payments, payment systems and intermediaries that facilitate noncash payments have the leads. Payment systems, you ask? They're the rules and processes that facilitate the exchange of goods and services for money. The systems use a series of electronic messages to move account-based money, as covered in Chapter 1.

Noncash payment systems and intermediaries are made up of large-value funds transfer systems, ACH operators, and credit and debit card systems. Together, these systems offer benefits that cash just can't: they're capable of moving large chunks of change across geographic distances, they're traceable, and they permit some payment automation and payment requests.

For example, the Fedwire Funds Service is "a real-time gross settlement system that enables participants to initiate funds transfer[s] that are immediate, final, and irrevocable once processed ... and is generally used to make large-value, time-critical payments."[14] Launched in July 2023, FedNow is the Federal Reserve's new instant payment service that lets customers at participating banks and credit unions send and receive money within seconds, 24/7, every day. Consumers, businesses, and nonbank payment providers can't use FedNow directly, though; they have to go through a participating financial institution.[15]

The Clearing House Interbank Payments System (CHIPS) is the largest private sector clearing system for transferring and settling payments in US dollars. CHIPS is privately operated by The Clearing House, which is owned by the approximately fifty financial institutions that participate in its system. CHIPS operates from 9 a.m. to 6 p.m. Eastern Time. After hours, CHIPS releases and nets unresolved payments and sends payment orders to Fedwire to settle.

Cross-border payments typically settle through correspondent banking networks. These are arrangements where one bank (correspondent) holds deposits owned by other banks (respondents) and provides payment and other services to those respondent banks.[16] This often requires a network or intermediary institution to ensure everything goes smoothly. An example is the Society for Worldwide Interbank Financial Telecommunication (SWIFT). SWIFT is a global financial messaging network that enables financial institutions worldwide to securely exchange information and electronic messages about financial transactions.

No matter the specific type of payment system, to properly function, all systems require the successful completion of two fundamental processes: clearing and settlement.

Clearing answers two payment-related questions: who's paying whom, and for what amount? Settlement involves the transfer of funds to discharge a debt. Regardless of the specific payment used, clearing and settlement always involve a payer (sender-debtor) and a payee (receiver-creditor).

For clearing and settlement purposes, payment systems require the following: (1) parties to the transaction and a network of participants—a sender, a receiver, and often one or more financial institutions; (2) an asset or set of assets that are transferred among those participants; and (3) a transfer process that defines the procedures and obligations associated with the transaction. They work to clear and settle the transaction. It's their job to resolve the respective debits and credits of the account-based money for the payer and payee.

Settlement risk is an important consideration in the world of payments. It's fraud or the possibility that a payer stiffs the payee (or vice versa). It matters because most payments aren't instant. (Rarely are people buying hockey tickets with cash today.) The majority of businesses have lines of credit and cash reserves to mitigate risk during the time it takes to send and receive payments. Both come with costs.

The Ways We Pay: From Barter to Tap-and-Go

Money and payments are inextricably linked. Think of the former as a means to the latter: money allows us to pay for food, shelter, and Netflix subscriptions. Before modern money existed, those same needs existed. (Well, maybe not the need to stream Netflix shows, but you get my point.)

This section provides a brief overview of the evolution of payments.

BARTER

We used to regularly—and inefficiently—exchange goods and services for other goods or services without exchanging money. In colonial times, barter typically involved tobacco, grain, and wampum, a traditional shell bead of the Eastern Woodlands tribes of Native Americans.

The rise of cash economies or monetary systems mostly put the nail in barters' coffins, but you can still find examples of it today. For example, I routinely visit my local farmer's market in New York City. I often observe vendors trading vegetables for milk, meat, or fish and vice versa.

CURRENCY

Many people conflate *money* and *currency*, but there's an important difference between the two. Money is a broad term. It refers to an intangible system of value that enables the efficient exchange of goods and services. Physical currency is one *form* of money—and presently the most popular one. (We'll see in Chapter 4 how digital currencies are gaining steam.)

Currency developed through tradable items, including animal skins, weapons, and even salt. Writing for NPR, Wim Hordijk notes:

> Being so valuable, soldiers in the Roman army were sometimes paid with salt instead of money. Their monthly allowance was called "salarium" ("sal" being the Latin word for salt). This Latin root can be recognized in the French word "salaire"—and it eventually made it into the English language as the word "salary."[17]

For a long time, we traced the first known coinage of money to the Anatolian Kingdom of Lydia in 630 BCE.[18] (I'll save you the trouble of googling. It's modern-day Turkey.) Recently research published in the journal *Antiquity*, however, casts doubt on that long-held belief. China may in fact have minted the first coins more than 2,600 years ago.[19]

Fast-forward to the eleventh century for the birth of paper money. As Jacob Goldstein wrote in his fascinating 2020 book *Money: The True Story of a Made-Up Thing*:

> The Mongols were nomads, and they loved how much easier it was to move paper money than metal coins. They understood that speed meant wealth. So the year Kublai became the Great Khan, he created a new kind of paper money, to be used across vast swaths of the empire. He called it the "inaugural treasure exchange voucher." (It's not just paper; it's a voucher you can exchange for treasure!) Kublai Khan really wanted people to use his new paper money, so he made it illegal to use bronze coins for trade.

The paper, which wasn't even pretending to be treasure vouchers or silver IOUs, still worked as money.

Marco Polo witnessed a radical experiment: money backed by nothing other than government authority. This was a testament, partly, to the sheer power of the Mongol state. It outlawed the use of bronze coins for trade. Marco Polo was immediately smitten with the banknotes around 1260. After seeing them in China, the Venetian merchant and explorer brought paper money to Europe.[20]

Of course, we now use plenty of non-paper, non-coin currencies to make payments. As Ludwig Von Mises writes in *The Theory of Money and Credit*:

> When an indirect exchange is transacted with the aid of money, it is not necessary for the money to change hands physically; a perfectly secure claim to an equivalent sum, payable on demand, may be transferred instead of the actual coins. In this by itself there is nothing remarkable or peculiar to money.

We often pay for goods and services using an increasing array of methods, discussed next.

CHECKS

A check is a paper document that allows your cousin Phil to deliver funds to his friend Arthur. It authorizes a bank or credit union to pull an amount of money out of a specific checking account. The financial institution then withdraws money from the account of the sender or drawee (Phil in this case) and either deposits it into the account of the receiver or payee (Arthur here) or gives them the equivalent cash. Dismiss checks if you like as relics of a bygone era, but 55 percent of Americans still wrote them in 2022. Nearly a quarter wrote one per month.[21]

DEBIT CARDS

If paying by check doesn't strike you as the acme of efficiency, you're correct.

In 1966, the Bank of Delaware launched a debit card pilot program as an alternative to carrying cash or a checkbook.[22] Unfortunately, no system connected merchants to banks outside of their states. As a result, debit card adoption didn't exactly take the US—or the world for that matter—by storm.

Usage picked up in the 1980s and 1990s as more and more ATMs started cropping up across the country. In 1990, roughly 300 million transactions used debit cards.[23]

CREDIT CARDS

The first credit card arrived in 1950, courtesy of Diners Club. The genesis of "the world's first multipurpose charge card" is an interesting one. The short sidebar that follows provides a quick synopsis.[24]

The Birth of the Credit Card

Businessman Frank McNamara frequently left his wallet at home while dining in New York City. Without cash one night to settle the bill, he called his wife to come to the rescue, as she did often, because McNamara had a habit of forgetting his wallet at home. Eventually, McNamara teamed up with Ralph Schneider to create a small cardboard card known today as a Diners Club Card. With it, members could dine on credit at a variety of local establishments. In 1981, Citigroup purchased Diners Club, and Discover Financial Services subsequently acquired it in 2008.

We now delve into the mechanics of credit card payments, which will help inform the value proposition for digital currency payments.

Say that I want to buy a new car, so I trek down to my local Jeep dealership. The new Cherokee meets all my needs and my budget. I ultimately decide to pull the trigger on it for $37,000. To keep things simple, I'm not trading in a car, and I'm paying full sticker price with no cash down. To initiate the transaction, I whip out my credit card.

We've now begun the multistage process that involves a number of players to complete the transaction. The transaction involves the cardholder (me), the merchant (the Jeep dealership), the acquiring bank (the dealership's bank),* the issuing bank (the bank that issued my credit card), and the card associations (Visa, Mastercard, American Express, Discover, etc.).

The transaction likely also involves a payment processor or payment gateway.

What's a Payment Gateway?

A payment gateway collects and verifies a customer's credit card information (name on card, card number, expiration date, and credit card verification (CCV)). It's particularly crucial for online payments, where there's no physical card reader. The gateway ensures that all the details of the transaction are correct so the sale information can be transmitted to the payment processor. The gateway authenticates and encrypts cardholder data, protecting it to relay the information from the merchant to the acquiring bank and then to the card issuer.

* An acquiring bank is a member of the card associations. It contracts with businesses to create and maintain accounts that allow the business to accept credit and debit cards. The acquiring bank also deposits funds from credit card sales into a merchant's account. Acquiring banks or payment processors provide equipment and software to accept cards and handle customer service and other necessary aspects involved in card acceptance.

A payment processor communicates transaction information between the merchant, the issuing bank, and the acquiring bank and initiates the transfer of funds. Examples are Stripe, Square, PayPal, Fiserv, Worldpay, Global Payments, and Clover.

The credit card transaction process involves the following steps:

- **Authorization and authentication:** The Jeep dealership swipes my credit card and transmits the information and the details of the transaction to its payment processor. The information is then routed to the issuing bank for approval through the card's network.* After confirmation that the transaction information is valid, that I have sufficient credit to make the purchase, and that my account is in good standing, the card issuer sends an approval response code to the acquiring bank or processor and to the merchant's terminal, software, or gateway.
- **Clearing and settlement:** The payment processor sends a batch of credit card transactions to the card network. The processor also deposits the funds from those transactions into the dealership's bank account and deducts processing fees.

Assuming that all is copacetic, I'll soon drive away with my new Jeep.†

* Mastercard transactions are routed through Mastercard's BankNet network; Visa transactions are routed through Visa's VisaNet network.

† See https://tinyurl.com/repay-credit for much more on the mechanics of credit cards.

Where Tech Meets Payments: The Rise of Fintech

A new breed of technology allows consumers to pay in even more ways. The World Bank defines financial technology (*fintech*) as "the application of digital technology to financial services." Think of it as the love child of banking and tech.

Scores of companies are developing and deploying powerful technologies to facilitate payments in interesting ways. The *New York Times* detailed the evolution of fintech in 2016.[25] Fintech is the "application of digital technology to financial services."[26] I've summarized a few of the most salient events in Table 2.1.*

Year	Event
1993	Citicorp establishes the Financial Services Technology Consortium with the goal of sponsoring collaborative tech-based research. It coins the term *financial technology*.
1994	The Japanese automotive company Denso Wave invents quick response codes. They languish for more than a decade, finally taking off with the advent of smartphones.†
1997	Amazon patents 1-Click payments, as discussed at the start of this chapter.
1998	Electronic-payments company Confinity (rebranded as PayPal) is founded to facilitate online payments.
2009	Twitter cofounder Jack Dorsey and American financial entrepreneur Jim McKelvey develop Square. The company's now-ubiquitous small credit card reader plugs into iPhones and allows physical merchants to easily accept customers' mobile payments. (In late 2021, the company rebranded as Block.)
2010	The payment service provider Stripe is founded. It accepts credit cards, transactions from digital wallets, and many other payment methods.

* For far more on the history and mechanics of payments, check out *The Pay Off: How Changing the Way We Pay Changes Everything.*

† See for yourself at https://tinyurl.com/magqrcode.

Year	Event
2011	Google launches Google Wallet. It allows consumers to use smartphones equipped with a near-field communication chip to make tap payments. Apple Pay followed in 2012.
2013	San Francisco-based Plaid launches. Its core technology enables applications to connect with users' bank accounts.
2015	The Chinese ecommerce giant Alibaba announces Smile-to-Pay, which enables consumers to authenticate mobile payments by scanning their face with their smartphones.

Table 2.1: Fintech Highlights Over the Past Three Decades

Scores of companies are developing and deploying powerful technologies to facilitate payments in innovative ways. Each innovation is different in its own way, but all share two common characteristics. First, they give consumers more choice over how they pay for goods and services. Second, they reduce the friction associated with payments.

This chapter has covered payment systems, without which modern businesses couldn't function. The next chapter delves deeper into the workings of one of the most essential of those institutions: the banking world.

Chapter Summary

- A payment is the delivery by the payer of money or other value to the payee to satisfy a debt.
- Smooth payments facilitate commerce. When they work effectively, economies run efficiently.
- Especially since COVID-19, the trend has been an acceleration away from cash to electronic forms of payment.

- One of the most popular methods of payment, the credit card, involves a multistage process and a number of players to complete a single transaction.
- Developing and deploying powerful technologies to facilitate more convenient, faster, and safer payments is an important trend.

Chapter 3

Banking: It's Complicated and Always Has Been

"A bank is a place that will lend you money if you can prove that you don't need it."

—BOB HOPE

In 1893, the US entered a four-year depression—the worst in its history. Blame a surplus of railroad building and the collapse of two of the nation's largest employers: the Philadelphia and Reading Railroad and the National Cordage Company.[1]

For good reason, Americans became increasingly concerned about their economic safety. Untold numbers rushed to withdraw their money from banks and caused bank runs.[2] A full-blown bank panic ensued, with many institutions suspending operations and others shuttering their doors for good. At this point in our story, you might be thinking: why didn't the Fed intervene?

Because it didn't yet exist. No government agency or entity could attempt to stabilize the economy via monetary policy. Instead, financial mogul J. P. Morgan intervened. He loaned the US Treasury $65 million in gold to preserve the gold standard and prevent economic collapse.[3] Yes, a single private citizen (albeit a very rich one) bailed out the US government. It's nothing short of astonishing, and his legacy remains. In 2008, J. P. Morgan & Co. bailed out Bear Stearns, and in 2023, it bailed out First Republic Bank.

Why do these events matter? It would be difficult to write a book about payments without devoting space to a key and adjacent topic: banking.

This chapter serves two purposes: it provides a brief overview of the American banking system, and it details the role of banks in causing one of the most calamitous economic events of our lifetimes: the 2008 global financial crisis.

The Evolution of Banks

The origins of banking stem from ancient Greek, Roman, Egyptian, and Babylonian civilizations. Early banks were effectively safe havens—temples that would deter thieves and other bad actors. As Benjamin Bromberg writes in *The Economic History Review:*

> Banking is one of the oldest institutions known to man. Its history is lost somewhere in remote antiquity. But this much is definitely known: banking was born in the temples consecrated to the gods and goddesses of mythology in the Mesopotamian area thousands of years before the rise of Christianity. This custom of sanctuary depositories spread in due time to

> the other ancient civilisations [*sic*] on the continent of Europe. And Rome was no exception."[4]

In the Middle Ages, there was, however, opposition to the idea of a holy institution charging fees for borrowing money:

> The collapse of trade after the fall of the Roman empire makes bankers less necessary than before, and their demise is hastened by the hostility of the Christian church to the charging of interest. Usury comes to seem morally offensive. One anonymous medieval author declares vividly that "a usurer is a bawd to his own money bags, taking a fee that they may engender together."[5]

Over time, politicians, philosophers, religious entities, and economists have largely agreed that, at a minimum, banks are necessary. Still, there was no unanimity about how banks should operate and what role they should play in society.

US BANKS: EARLY INFLUENCERS

As international trade increased in the 1700s, Adam Smith advocated for central banking in his opus *The Wealth of Nations*. Money, coinage, and payments were necessary conditions for markets to function properly and efficiently.

Way before the Broadway debut, Alexander Hamilton served as the first US Secretary of the Treasury from 1789 to 1795. With the country facing inflation and crippling post-war debt, he advocated for creating a national bank. Using the charter of the Bank of England as a model, Hamilton proposed a financial institution that could do each of the following:

- Issue paper money (whether in banknotes or currency).
- Safely store public funds.
- Offer banking facilities for commercial transactions.
- Act as the government's fiscal agent, including collecting the government's tax revenues and paying the government's debts.

Fellow Founding Father Thomas Jefferson disagreed. He viewed banking as a state activity and the idea of a national bank as unconstitutional.[6] He argued that it violated the Fifth Amendment of the Constitution, which states, "The powers not delegated to the United States by the Constitution, nor prohibited by it to the States, are reserved to the States respectively, or to the people."[7]

Hamilton's view ultimately prevailed. The Bank of the United States, the country's first national central bank, opened in Philadelphia on December 12, 1791. Its charter expired in 1811, and Congress did *not* renew it. The economic turmoil of the War of 1812 and the Civil War led to the creation of several other national banks in the nineteenth century.

As the new century approached, the US was rapidly industrializing and, not coincidentally, urbanization was on the rise. As a result, merchant banks started gaining power and popularity. Goldman Sachs, Loeb & Co., J. P. Morgan & Co., and their ilk dealt primarily in commercial loans and investments; they didn't provide traditional banking services to everyday Americans.

THE CREATION OF THE FEDERAL RESERVE

Eager to avoid a repeat of the bank run and financial instability mentioned at the start of this chapter, in 1913, Congress passed the Federal Reserve Act. President Woodrow Wilson signed it into law. The legislation sought to "enhance the stability of the American banking system"[8] by creating the Federal Reserve Bank (Fed). The Fed consisted of twelve branch banks. The legislation also created *Federal Reserve notes*: US dollars backed by the Fed's assets. Previously, each commercial bank issued its own notes.

The Fed not only created a new, unified national currency, but it made the processing of payments faster and more efficient. The twelve Federal Reserve Banks provide check-clearing services for one another. An early innovation was the development of an electronic system for making long-distance payments using the telegraph, which later became known as Fedwire.[9]

THE MODERN-DAY BANK

Today, banks provide core economic functions. First, they offer financial intermediation such as lending and investing the money we deposit with them. Banks also (critically) supply credit to enterprises, households, and governments. This requires the operation of key payments systems, as described in Chapter 1.

Our core banking system has actually remained *relatively* unchanged over the past century, at least to the average citizen. Behind the scenes, though, it's unrecognizable from a century ago—thanks to technology. As previously discussed, digital innovations have transformed how financial institutions serve their customers and move their money around.

With respect to banks, consider the following:

- Mainframe computers in the 1960s.
- Credit card and ATM networks in the 1970s.
- Electronic trading of stocks and equities in the 1980s.
- Online banking in the 1990s.
- Mobile banking in the 2000s.

The Banking System Works—Until It Doesn't

After the dot-com crash, the Fed deliberately kept interest rates historically—and some would argue, *artificially*—low. As Figure 3.1 demonstrates, they hovered near zero for years.

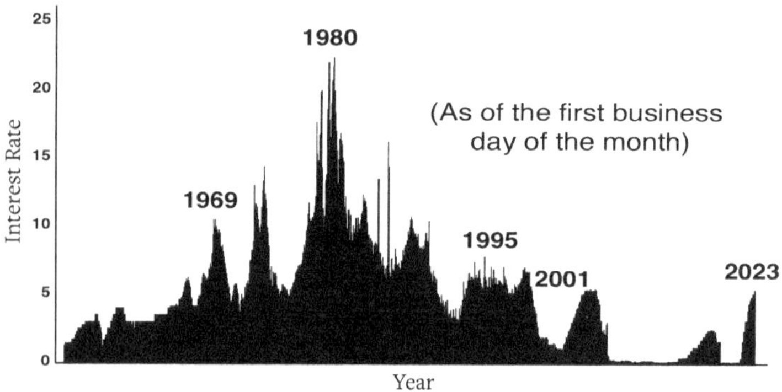

Figure 3.1: Federal Funds Rate Since 1954*
Source: Macrotrends https://tinyurl.com/mgint-2a

Cheap money made obtaining the American dream of home ownership more attainable. Millions of Americans flocked to buy their first homes or refinance their existing mortgages.

Within a few years, the conventional mortgage market became saturated. Bankers faced a daunting reality: an inability to sell new business. Specifically, they were unable to underwrite as

* The *Federal Funds Rate* is the interest rate at which depository institutions trade federal funds (balances held at Federal Reserve Banks) with each other overnight.

many safe, 30-year fixed mortgages to first-time buyers with high FICO scores. So banks got creative.

THE GREAT MODERATION: THE BEST OF TIMES

The mid-1980s to 2007 was a period of relative macroeconomic stability that's known as the Great Moderation. Under the leadership of Federal Reserve Board Chairs Paul Volcker (ending in 1987), Alan Greenspan (1987–2006), and Benjamin Bernanke (starting in 2006), inflation was low and stable. The US saw the longest period of economic expansion since World War II.[10]

The Federal Reserve cut interest rates early in the new century. In turn, mortgage rates fell, causing a surge in home refinancing. Everyday Americans could withdraw equity that was built up over previous decades in their homes and spend more money, despite stagnant wages. Home sales volumes and prices surged. Cash became cheap, and the thirty-year fixed-rate mortgage, with a 20 percent down payment, went out of style.

Banks and other lenders created new products such as subprime and adjustable mortgages. We also saw the advent of NINJA (no income, no job, and no assets) loans, underwritten based on a borrower's credit digits with no verification of how much dough they were bringing in or what they owned. NINJAs gave a fair shake to borrowers who may not have otherwise qualified for much in the way of lending. The expectation, however, was that interest rates would remain low and home prices would continue to rise.

The terms of these loans were full of legalese, and whether borrowers actually read—much less understood—the complex fine print is debatable. In 2005, the American Bar Association criticized banks for engaging in predatory lending, which they asserted was “the new face of economic injustice.”[11] The ABA wrote,

> As fringe lenders become a pervasive presence in low-wealth neighborhoods, economic justice concerns have shifted away from access and to the terms of credit. White borrowers tend to be served by banks and other conventional institutions in the prime market. In contrast, people of color, women, and the elderly are targeted by high-cost lenders.[12]

Many unsuspecting borrowers would ultimately default once the teaser rates on their subprime loans expired or simply because they never had the ability to repay the loan in the first place. To mitigate this risk, lenders used a complex array of financial products, including mortgage-backed securities, collateralized debt obligations, and credit default swaps.

Researchers at the Institute for Research on Labor and Employment at the University of California, Berkeley, studied the widespread adoption of these financial products. In their words, the financial industry:

> began to bundle lower-quality mortgages—often subprime mortgage loans—in order to keep generating profits from fees. By 2006, more than half of the largest financial firms in the country were involved in the nonconventional [mortgage-backed security] MBS market. About 45 percent of the largest firms had a large market share in three or four non-conventional loan market functions (originating, underwriting, MBS issuance, and servicing). By 2007, nearly all originated mortgages (both conventional and subprime) were securitized.[13]

The numbers are flabbergasting. In 1995, only 30 percent of all subprime loans were securitized; by 2000, it was 40 percent; and by 2007 it was 90 percent.[14]

SEPTEMBER 2008: THE WORST OF TIMES

2007 and 2008 brought the greatest financial crisis the US had seen since the Great Depression. The collapse of the housing market—fueled by low interest rates, easy credit, and predatory lending practices—led to the 2008 economic crisis. New home sales fell by a record amount in 2007.[15] Ultimately, the bust of the housing market bubble spilled over into the entire financial services sector. Financial institutions were highly leveraged and had inadequate capital reserves, among other things. This made them extraordinarily vulnerable to the downturn in the market.

Bear Stearns collapsed on March 16, 2008, after the Federal Reserve Bank of New York denied a cash loan of $25 billion so that it could stay in business. J. P. Morgan purchased Bear Stearns for pennies on the dollar.[16] Lehman Brothers, which by 2007 was the world's largest holder of mortgage-backed securities, had incurred $619 billion in debt.[17] Its bankruptcy filing in September 2008 was the largest corporate bankruptcy filing in US history.[18]

The Lehman Brothers catastrophe rattled the markets. Other dominoes started falling—and fast. Both on Wall Street and Main Street, a sense of foreboding prevailed. (As a Manhattan resident at the time, I can attest to the general consternation. It was downright scary.) Amid the chaos and uncertainty, there was one flicker of light: at least it wasn't 1893. The Federal Reserve possessed formidable arrows in its quiver to try to stabilize the economy.

On September 16, 2008, a day after Lehman Brothers filed for bankruptcy, the Federal Reserve provided an $85 billion two-year loan to American International Group to prevent its bankruptcy.[19] AIG, through an unregulated subsidiary, had insured counterparties against losses on certain mortgage-related financial products and didn't have the funds to perform its obligations.

A positive step to be sure, but it didn't staunch the bleeding.

The pressure to act—meaningfully—rapidly escalated. Federal Reserve Chair Ben Bernanke, US Treasury Secretary Henry Paulson, and Federal Reserve Bank of New York President Timothy Geithner all rushed to Congress to lobby for a $700-billion bill to bail out the banks.[20] (To be fair, the Dodd-Frank Act passed in July 2010, eventually dropping that staggering amount to "only" $475 billion.) Figure 3.2 displays the Troubled Asset Relief Program (TARP) allocations.[21]

By the time President George W. Bush signed the Emergency Economic Stabilization Act of 2008 on October 2, the financial contagion had spread far beyond the US. The Global Financial Crisis of 2008 had arrived in earnest. The ensuing fallout resulted in the highest rates of unemployment and home foreclosures in the US since the Great Depression. The latter mushroomed 81 percent from the prior year.[22]

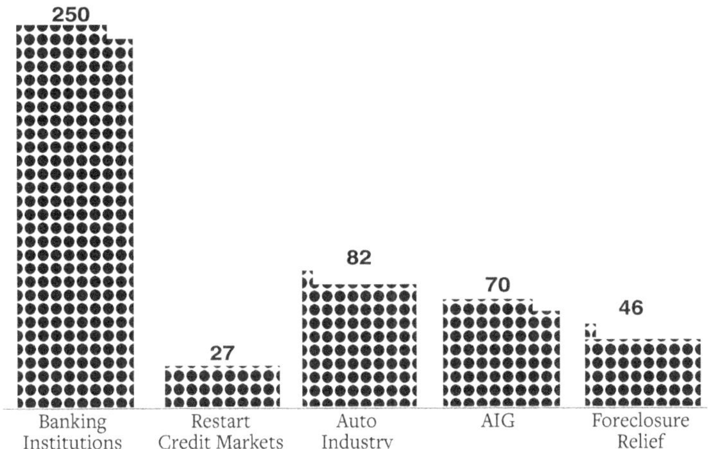

Figure 3.2: TARP Allocations (In Billions): Where the Money Went

Asleep at the Wheel?

Equipped with 20/20 hindsight, one could argue that the 2008 Global Financial Crisis never should have happened. Congress created the Financial Crisis Inquiry Commission to "examine the causes of the current financial and economic crisis in the United States." It produced a report with the results of its examination and its conclusions about the causes of the crisis.[23]

The Commission found that the financial crisis was avoidable because "[t]he captains of finance and the public stewards of our financial system ignored warnings and failed to question, understand, and manage evolving risks within a system essential to the well-being of the American public."[24] Decades of deregulation and reliance on self-regulation by financial institutions eliminated safeguards that could have helped avoid the crisis. Moreover, the Commission found that failures of corporate governance and risk management as well as a breakdown in accountability and ethics at many important financial institutions were key causes of the crisis. Excessive borrowing, risky investments, and a lack of transparency coalesced to create vulnerabilities for financial distress. And US regulators such as the Treasury Department, the Federal Reserve Board, and the Federal Reserve Bank of New York, which should have been in the best position to oversee markets, weren't prepared for the events of 2007 and 2008. A lack of mortgage-lending standards and failures of credit rating agencies were also essential to economic destruction.

THE AFTERMATH

The Great Recession technically ended in June 2009, but economic weakness persisted for years. Consumer confidence in the federal government and banks suffered. Gallup reported in 2004 that only about half of all Americans expressed faith in banks as institutions. By 2009, that number had plunged to 22 percent.[25] As of this writing, that number has barely budged, with Republicans, Democrats, and Independents all in general agreement.[26]

While the dust was still settling from economic collapse, in the summer of 2009, Citigroup, Goldman Sachs, and other remaining investment banks lavished exorbitant bonuses on their employees. More than 738 Citigroup employees received annual payouts exceeding $1M that year.[27]

As we'll see in the next chapter, the antipathy toward powerful, centralized institutions spawned at least one unexpected development: the creation of a decentralized payment system and a digital alternative to government-issued fiat currency.

Is our financial system safer now than it was in 2008?

In theory, the answer should be *yes*. After all, the Dodd-Frank Wall Street Reform and Consumer Protection Act of 2010 should have shored up the system. The legislation was passed to prevent a repeat of the 2008 Global Financial Crisis.[28] However, in 2018, as a result of pressure from the banking industry, the Trump Administration made significant changes that increased the threshold for increased scrutiny of banks from $50 billion in assets to $250 billion in assets.[29] With this foreshadowing, we head to the next section of this chapter.

Central Banking and a Forthcoming Recession?

I began writing this book in the summer of 2022. At the time, many mainstream economists believed that a recession was imminent.[30] Critical questions included whether the downturn would resemble 2008's Great Recession and how long it would last.[31]

Among other things, recessionary concerns at the time stemmed from the following:

- A decrease in consumer spending.
- Supply chain disruptions.
- General market volatility.
- An increase in public and private debt.
- Russia's invasion of Ukraine and the resultant spikes in energy, oil and natural gas, food, fertilizer, and industrial metals prices.

Nouriel Roubini is a professor emeritus at New York University's Stern School of Business. What's more, he was one of the first experts to forecast the 2008 Global Financial Crisis. In his words:

> Today, we face supply shocks in a context of much higher debt levels, implying that we are heading for a combination of 1970s-style stagflation and 2008-style debt crises—that is, a stagflationary debt crisis.[32]

This brings us back to central banks. After all, they oversee monetary policy. They can flex their muscles to prevent or mitigate the effects of a recession, but what are the downsides? Do central banks already possess too much financial power today?

Writing for the *Wall Street Journal* in October 2022, Judy Shelton argues that no other institution possesses "the power to create money with no questions asked, manipulate the cost of capital, or counteract movements in financial markets. The central bankers are in charge—and perhaps that should change.[33]

Against this backdrop, is another decentralized form of financial services not only possible but even advantageous and desirable? As we'll see in Part II, the answer is *yes*—at least when it comes to payments.

Out of the Woods?

By the fall of 2023, as I was putting the final touches on this book, inflation had dropped, and the sentiment among many economists had shifted.[34] The Fed appeared to be on the verge of engineering a soft landing.

We're not out of the woods, though. The Israeli-Hamas War, the continuation of student loan payments after the 3 1/2-year pandemic freeze, the auto strike, high borrowing costs, and frequent near-government shutdowns foreshadow another recession.[35]

Only time will tell whether the US will have to fend against another recession.

Chapter Summary

- For a long time, the US lacked both a national currency and a central banking system subject to federal governmental oversight.
- The advent of the Federal Reserve centralized US banking, created a national currency, and made processing payments faster and more efficient.
- The 2008 Global Financial Crisis demonstrated that large financial institutions can and do fail, particularly as a result of poor corporate governance and risk management controls.
- The 2008 Global Financial Crisis and the 2023 economic slowdown also showed the limits of the federal government's ability to manage monetary risk and caused an antipathy toward powerful, centralized institutions.

Chapter 4

Unpacking Digital Assets

"Sometimes it does happen that the good thing is uncomfortably closer to the bad thing than we would like."

—VITALIK BUTERIN

At the end of 2008, the global economy was teetering. Banks and other financial institutions were failing across the world. Cheap credit and lax lending standards caused crises to erupt far beyond the US. Those in Greece and Iceland were particularly acute. Against this ominous backdrop, wheels were in motion to turn the traditional notions of banking, currency, and money on their heads.

What if there were a fundamentally different and *decentralized* way to conduct financial transactions? Such a way wouldn't require the intervention of central banks, financial institutions, and other trusted third parties. Rather, a new technology would replace these essential functions.

This chapter explores the fascinating world of digital assets.

Disclaimers

Before proceeding, a few disclaimers are in order. Throughout Chapters 1 through 3, I frequently used the term *cryptocurrency*, though I never defined it. That was deliberate. Cryptocurrencies are just one type of digital asset, as I explain here. However, the term *cryptocurrency* or *crypto* has become generally synonymous with the entire asset class.

There has been a great deal of buzz about digital assets. Ditto for the level of confusion. It's not hard to find differing opinions: crypto projects and companies, financial institutions, academics, think tanks, and government agencies oftentimes disagree on digital assets' specific definitions, benefits, and risks. Their disparities are often more ideological and fundamental than semantic. In addition, there's a lack of uniformity in term meanings.

I'll be the first to admit that digital assets can be a bit heady and tough to digest. When I first dove into the space in 2015, it took me a while to wrap my head around all their complexities, interdependencies, and distinctions. Unlike today, there were few resources at the time to clear my confusion.

With these disclaimers out of the way, let's jump in.

The Birth of Bitcoin

Satoshi Nakamoto is an essential part of this story, but we actually know little about this mysterious person or *persons*.* On Halloween 2008, as the world was reeling from the 2008 financial crisis, Nakamoto published a now-famous white paper titled "Bitcoin: A Peer-to-Peer Electronic Cash System."[1] (I'll refer to this as the "Bitcoin white paper.") They correctly wrote that

* As a result, we'll go with the pronoun *they*.

"commerce on the Internet has come to rely almost exclusively on financial institutions serving as trusted third parties to process electronic payments." The massive problems of the day stemmed from this very system.

Rather than merely complaining about the inherent flaws of centralized banking, Nakamoto acted. The Bitcoin white paper created a "framework of coins made from digital signatures" and a "system for electronic transactions without relying on trust."*

> On or about January 3, 2009, Nakamoto mined the first fifty Bitcoins and started the Bitcoin payment network. Not long after that, on January 12, Hal Finney was the recipient of the first Bitcoin transaction.[2]

Odds are that Bitcoin is the first digital currency or electronic money form you heard about. Let me be clear, though: several others beat Nakamoto to the punch.

In 1983, the American computer scientist and cryptographer David Chaum conceived an anonymous cryptographic electronic money, believed to be the outline of the first cryptocurrency.[3] In 1989, Chaum started the company DigiCash (with eCash as its trademark), a digital-based system that facilitated the process of transferring funds anonymously.[4] Other early entrants included b-money (created by Wai Dai in 1998 as an anonymous, distributed electronic cash system) and Bit Gold (created by Nick Szabo in 1998 but never deployed, though it has been called "a direct precursor to the Bitcoin architecture").[5] For reasons that will become apparent shortly, these digital forms of currency never took off.

---

* How Bitcoin works is beyond the scope of this book. If you're interested in more information, I recommend visiting https://bitcoin.org/en/.

THE DOUBLE-SPEND PROBLEM

Let's take a step back for a moment. Say that I take my daughters to the local farmers market. I buy a few heads of farm-grown organic broccoli by handing the vendor a $10 bill. As we saw in Chapter 2, this simple act of payment discharges my debt to the farmer. And because the farmer received physical money, there's no concern that I can use that same $10 bill in another transaction because I don't have it anymore. In other words, physical money can be spent only once.

However, digital money can be duplicated. Historically, it has been difficult to prevent a person, machine, or organization from duplicating *any* digital content. (Ask execs from the record and entertainment industries if they've been able to solve the piracy problem over the past quarter-century. The answer is no.)

Hence, the "double-spend problem": if I receive digital money, how can I be sure that the sender didn't simultaneously send the money to someone else? Today we generally rely on trusted central authorities to prevent this double-spend problem and other types of fraud.[6]

As we saw in Chapter 2, traditional payments systems are built on trust and identity. We trust that the banks, payment processors, governments, and other third parties involved in each transaction will ensure its successful completion. Via trust and identity, the system generally works—even if friction and considerable transaction fees leave a bad taste in our mouths.

Compared to its predecessors, Bitcoin emanated from a starkly different and more skeptical financial environment. In a stroke of genius, Nakamoto cracked the double-spend nut, but in a *trustless* manner. They created the first truly decentralized digital form of money by turning to a nascent technology.

BLOCKCHAIN

Bitcoin is built on blockchain technology. A blockchain is a shared immutable ledger across a network of participants (referred to as *nodes*), where up-to-date information is available to all participants at the same time. All network participants have access to the distributed ledger and its immutable record of transactions. Transactions are recorded only once, eliminating the duplication of effort that's typical of conventional business networks. As each transaction occurs, it's recorded as a "block" of data. Each block is connected to the ones before and after it, forming a chain of data. Transactions are linked in an irreversible chain: a blockchain.*

> The idea of blockchain technology was introduced in 1991 by Stuart Haber and W. Scott Stornetta in their paper "How to Time-Stamp a Digital Document."[7] In this paper, they explained the use of a continuous chain of timestamps to record information securely.

If all of this seems a bit abstract, consider how Bitcoin transactions compare to their older, slower, trust-based alternatives. As Marco Iansiti and Karim R. Lakhani wrote for *Harvard Business Review* in 2018:

---

* A more detailed explanation of how blockchains work is beyond the scope of this book. For more reading on this topic, I recommend visiting https://online.stanford.edu/how-does-blockchain-work.

> … a typical stock transaction can be executed within microseconds, often without human intervention. However, the settlement—the ownership transfer of the stock—can take as long as a week. That's because the parties have no access to each other's ledgers and can't automatically verify that the assets are in fact owned and can be transferred. Instead a series of intermediaries act as guarantors of assets as the record of the transaction traverses organizations and the ledgers are individually updated.
>
> In a blockchain system, the ledger is replicated in a large number of identical databases, each hosted and maintained by an interested party. When changes are entered in one copy, all the other copies are simultaneously updated. So as transactions occur, records of the value and assets exchanged are permanently entered in all ledgers. There is no need for third-party intermediaries to verify or transfer ownership. If a stock transaction took place on a blockchain-based system, it would be settled within seconds, securely and verifiably. (The infamous hacks that have hit bitcoin exchanges exposed weaknesses not in the blockchain itself but in separate systems linked to parties using the blockchain.)[8]

Blockchains rely on cryptography, the mathematical and computational practice of encoding and decoding data. Cryptography ensures the security of the transactions and the participants, eliminates the need for a central authority to perform these functions, and protects against double-spending. (Chapter 6 will return to this subject in the context of digital wallets.)

Since the advent of the Bitcoin blockchain, many others have followed.

We've already covered Bitcoin and the Bitcoin blockchain in some detail. I'd be remiss not to include a brief section on the Ethereum network, the foundation for the second most valuable digital currency, Ethereum.*

The Bitcoin blockchain is powerful but relatively limited: it only allows participants to transfer Bitcoin to one another. It also takes a relatively long time to process transactions on the Bitcoin blockchain. By contrast, the Ethereum blockchain represents much more. It's an open-source application that makes all sorts of exciting things possible, including:

- **Smart contracts:** These are "simply programs stored on a blockchain that run when predetermined conditions are met."[9] They're self-executing. Think of them as vending machines: you automatically receive a can of soda or a snack after you insert your money and make a selection.
- **Decentralized applications (dApps):** These are applications built on a decentralized network that combine smart contracts and front-end user interfaces.[10] They enable specific use cases for blockchain technology, such as trading, gaming, and staking.
- **Decentralized finance (DeFi):** A term used to refer to financial services and products that are available and accessible to anyone who can make use of Ethereum or other distributed ledger computing platforms.
- **Non-fungible tokens:** NFTs are tokens that can be used to represent ownership of unique items like art, collectibles,

* At least according to coinmarketcap.com.

> and real estate. Ownership of the asset is secured by blockchain—no one can modify the record of ownership or copy/paste a new NFT into existence. They're minted through smart contracts that assign ownership and manage the transferability of the asset through the unique ID and metadata that no other token can replicate.*

An important note: one must distinguish a blockchain (the Bitcoin or Ethereum blockchain) from the native asset that secures it.

Some other blockchains to be aware of follow:

- **Algorand:** Founded in 2017, it's a blockchain platform designed to process payment transactions quickly. Like Ethereum, it provides infrastructure to support the development of other blockchain-based projects like smart contracts.†
- **Avalanche:** Founded in 2020 by researchers from Cornell University led by Emin Gün Sirer, Maofan "Ted" Yin, and Kevin Sekniqi, it's another blockchain platform that competes with Ethereum. The Avalanche blockchain promotes "near-instant transaction finality."‡

---

* Fun fact: in March 2021, Mike Winkelmann, the digital artist known as Beeple, fetched a mind-boggling $69 million for his NFT "Everydays—The First 5000 Days." Before that windfall, the most he had ever sold a print for was $10.

† For more information, check out www.algorand.foundation/.

‡ For more information, check out www.avax.network/.

- **Litecoin:** Founded in 2011 by a former Google engineer named Charlie Lee, it was created based on Bitcoin's source code as an alternative to Bitcoin to, among other things, process transactions faster than the Bitcoin blockchain.*
- **Solana:** Also founded in 2017, it's a blockchain platform designed to host decentralized, scalable applications. It has smart contract capabilities, can be used for decentralized finance (DeFi) transactions, and can create NFTs. It promotes faster and cheaper transactions than rival blockchains like Ethereum.†

Blockchain is a transformative technology with implications for healthcare, supply chain, and government. All are beyond the scope of this book, however.

Now it's time to dig deeper into digital assets.

Explaining Digital Assets: Definitions and Classifications

For the purposes of this book, a *digital asset* is an umbrella term that encompasses any intangible asset that's created and stored in digital format. The digital asset could be secured on blockchain, distributed ledger, or any similar technology, but it doesn't have to be for purposes of this definition. For example, loyalty rewards such as American Express, Starbucks, and Barnes & Noble points are digital assets. This section covers the following categories of digital assets:

* For more information, check out https://litecoin.org/.

† For more information, check out https://solana.org/.

- **Digital currencies:** These are cryptocurrencies, digital tokens, price-based stablecoins, and central bank digital currencies.
- **Other blockchain-based tokens:** Examples include NFTs, security tokens, and utility or other tokens.
- **Digital loyalty points:** These encourage customers to spend money at a brand's stores and other places of business.

Figure 4.1 displays a simple visual of this vast universe.

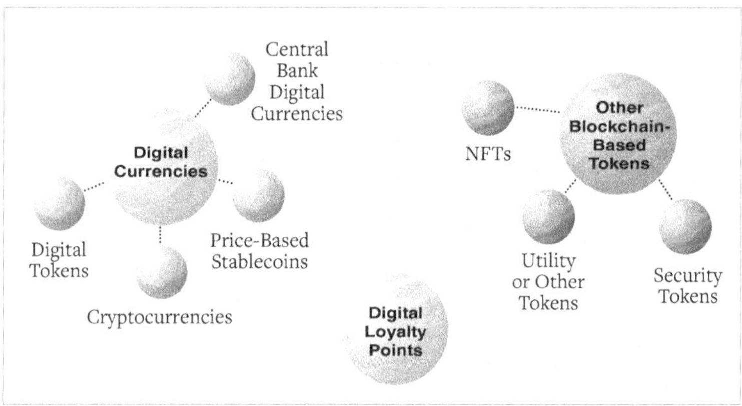

Figure 4.1: The Digital Asset Universe

Now that we've listed the types of digital assets, we'll unpack each one.

DIGITAL CURRENCIES

The largest category discussed here, and the focus of this book, is *digital currencies*. Digital currencies are blockchain-based assets that can be used as a medium of exchange. Digital currencies include cryptocurrencies, digital tokens, price-based stablecoins, and central bank digital currencies. A natural question at this point is, how many digital currencies exist today? It's a simple question, but the answer is tough to accurately gauge. The best estimate is that there are about 23,000[11] as of this writing.

Cryptocurrencies

Cryptocurrencies are the native asset that secures a particular blockchain. They're created and issued directly by the blockchain protocol they run on. For example:

- **ALGO:** The native cryptocurrency of the Algorand platform. It's used to secure the Algorand blockchain and pay processing fees for Algorand-based transactions.
- **AVAX:** The native cryptocurrency of the Avalanche blockchain. It's used to pay transaction processing fees, secure the Avalanche network, and act as a basic unit of account among chains in the Avalanche network. (AVAX is pronounced "uh-VOX" in case you're curious.)
- **ETH:** The native currency of the Ethereum blockchain. It's used to pay transaction fees on the Ethereum blockchain.
- **LTC:** The native currency of the Litecoin blockchain.
- **SOL:** The native cryptocurrency of the Solana blockchain. It's used to secure the Solana blockchain and to pay processing fees for Solana-based transactions.

Digitial Tokens

Digital tokens utilize and are built on an existing blockchain, such as Ethereum. These tokens can serve a multitude of functions, including for participating in decentralized finance (DeFi), accessing platform-specific services, and even playing video games. Some examples include:

- **COMP:** Created by Compound, a decentralized, blockchain-based protocol that allows people to lend and borrow cryptocurrency. The COMP token empowers community governance of the Compound protocol; COMP token-holders and their delegates debate, propose, and vote on all changes to the protocol.*
- **LINK:** Created by Chainlink, a technology platform that enables non-blockchain enterprises to connect blockchain-based smart contracts with external data, such as baseball scores or stock prices. The LINK token facilitates transactions on the platform.
- **STX:** Created by Stacks, an open-source blockchain network that brings decentralized apps (dApps) and smart contracts to Bitcoin. It's the network's native cryptocurrency used on the Stacks network to pay transaction fees and deploy smart contracts.

Price-Based Stablecoins

These stablecoins are designed to maintain price stability relative to a government currency.† In the US, stablecoins primarily facilitate payments, trading, lending, and borrowing of other digital assets. They can support faster, more efficient, and more inclusive payments options. While their underlying collateral structures vary, price-based stablecoins aim for the same goal: price stability.

---

* For more information, check out https://compound.finance/.

† Some websites and institutions define *stablecoins* as types of cryptocurrencies. I don't in this book because they're not the native currency of a particular blockchain. Instead, they're better defined as a type of digital token.

Price-based stablecoins generally rely upon two structures. Table 4.1 displays them.

Type	Description
Fiat-backed	This stablecoin structure is pegged to the value of—and generally collateralized with—an underlying government-issued fiat currency. The ratio is usually 1:1. This type of stablecoin takes advantage of the stability provided by central banks and the government that issued the underlying fiat currency. In theory, this should reduce a fiat-backed stablecoin's volatility. Two of the most popular fiat-backed stablecoins are USD Coin and Pax Dollar. Both stablecoins are pegged to the United States dollar. In July 2023, PayPal entered the stablecoin arena with PYUSD.
Fiat-pegged	This stablecoin structure relies on the use of specialized algorithms and smart contracts to manage the supply of tokens in circulation to stabilize prices. For example, an algorithmic stablecoin system reduces the number of tokens in circulation when the market price falls below the price of the fiat currency it tracks. Dai is a popular algorithmic stablecoin pegged to the US dollar. It implements game theory to stabilize the price of the token. Whenever the value of Dai goes below $1, the protocol incentivizes users to increase the price. In contrast, if and when the value of Dai goes above $1, the incentives work the other way around. TerraUSD (UST) was another popular algorithmic stablecoin pegged to the US dollar until it experienced destabilization in May 2022.[12] (Chapter 9 returns to this subject.)

Table 4.1: Price-Based Stablecoin Underlying Collateralization Structures

Central Bank Digital Currencies

Central bank digital currencies (CBDCs) represent a digital version of a nation's fiat currency. As such, central banks issue and collateralize them. Think of a CBDC as a digital form of paper money that the government makes available to the general public.

Federal Reserve policymakers and staff have studied CBDCs closely for several years. The Biden administration is wisely exploring the implications of—and options for—issuing a digital dollar. The Digital Dollar Project (DDP) is an exciting partnership between Accenture and the Digital Dollar Foundation. It aims to advance the exploration of a US CBDC.*

OTHER BLOCKCHAIN-BASED TOKENS

This section discusses other types of tokens that are built using blockchain technology. However, these tokens would generally not be used as a medium of exchange in a payment transaction. This is why they're not categorized as digital currencies. There are many types of blockchain-based tokens. A few significant ones deserve mentioning here:

- **Non-fungible tokens (NFTs):** Mentioned earlier in this chapter, these tokens represent a number of things, such as art, music, in-game items, tickets, and collectibles in both the physical and the digital realms. NFTs allow the holder to own an original item of a limited supply, originality, or edition. They've helped artists, creators, and collectors sell their items and receive royalty payments for sales whenever the content is sold to a new user. In early 2023, Sesame Workshop introduced the first digital collectibles based on the iconic Sesame Street kids' brand. It featured 5,555 editions of a digital Cookie Monster for sale, priced at $60 each.[13] Another example is *CryptoKitties*, one of the

* For more information, check out https://digitaldollarproject.org/.

first blockchain video games that allowed users to breed and collect digital cats.

- **Security tokens:** These tokens represent ownership of an asset, such as equity in a company, and can be used to raise capital. For example, BCAP, an Ethereum-based smart contract token, was the first tokenized venture fund by Blockchain Capital.
- **Utility and other tokens:** These tokens generally provide holders with access to specific products and services and are sometimes distributed during crowdfunding sales. An example is ApeCoin (APE), which enables holders to participate in the Bored Ape Yacht Club ecosystem's governance votes and to access members-only features like games, events, and other services. Another example is Decentraland (MANA), a payment for land and other goods in the virtual gaming world of Decentraland.

DIGITAL LOYALTY POINTS

The final type of digital asset is digital loyalty points. Brands issue them to encourage customers to spend money at their stores and other places of business. The more consumers spend with the brand, the more points they can earn via these loyalty programs. Individuals can usually exchange loyalty points for products, upgraded services, and enhanced experiences.

Loyalty points and programs are no longer the sole purview of airlines and credit card companies. For example, Chipotle launched its loyalty program in 2019, and the results have been spectacular. As of this writing, more than 27 million members participate.[14]

Successes aside, though, most legacy loyalty programs can up their games. Different point systems, byzantine rules, and limited redemption options add considerable friction to redeeming and exchanging points. If you think blockchain can solve this problem, trust your judgment.[15]

In early 2022, Shake Shack ran a promotion in which customers received Bitcoin as a reward for purchases made at the brand using Block's (SQ) Cash App.[16] For its part, the Brazilian fintech bank Nubank announced the creation of the Nucoin token on the Polygon blockchain in October 2022. The token paved the way for a successful rewards program for tens of millions of its clients across Latin America.[17]

Crypto Criminals

This chapter has explained the many uses of digital assets and has defined *digital currencies* for further elucidation in Parts II and III of the book. As we'll see in Part II, the potential is massive when it comes to digital currency payments. Unfortunately, though, digital assets, like cash and diamonds, can usher in illicit activities.

SILK ROAD

You can buy just about any legal good or service online, but what about illegal ones?

Libertarian and cyberpunk Ross William Ulbricht likely asked himself that question many times. Ultimately, he decided to do something about it. In 2011, he started Silk Road, the Amazon of the Dark Web. To preserve his anonymity (or so he thought), Ulbricht went by the name Dread Pirate Roberts, a nod to the fictional character in *The Princess Bride*.

Looking for prostitutes, hitmen, poison, methamphetamine, or industrial-grade weapons? Silk Road had you covered as long as you could pay in Bitcoin, which purportedly aided in protecting user identities.*

Silk Road quickly took off—and it didn't take long for the authorities to notice. In January 2012, the Department of Homeland Security arrested a Silk Road user called DigitalInk. That individual ultimately decided to cooperate with authorities, and the chase for Ulbricht was on. (Nick Bilton's bestselling 2018 book *American Kingpin: The Epic Hunt for the Criminal Mastermind behind the Silk Road* is one of the most compelling reads of the past decade.)

By October 2013, the FBI started tracking Ulbricht. The feds arrested and indicted him on charges of narcotics conspiracy, money laundering, and solicitation of murder for hire. He's currently serving a life sentence.

COLONIAL PIPELINE RANSOMWARE ATTACK

In May 2021, hackers brought a major gas pipeline to a standstill demanding ransom to restore operations. In case you're unfamiliar with the Colonial Pipeline (as I was before researching this book), it's one of the largest and most essential oil pipelines in the US. It was built in 1962 to help transport oil from the Gulf of Mexico to the East Coast states.

The US government classified it as a national security threat, and President Joe Biden declared a state of emergency.

---

* In case you're curious, the name *Silk Road* comes from a historical network of Eurasian trade routes started during the Han Dynasty (206 BCE–220 CE) between Europe, India, China, and many other countries on the Afro-Eurasian landmass.

Colonial Pipeline acceded and paid 75 Bitcoin for ransom, but the story doesn't end there. One month later:

> The DOJ was able to find the digital address of the wallet that the attackers used and got a court order to seize the bitcoin. The operation recovered 64 of the 75 bitcoin that Colonial Pipeline paid. At the time of the recovery, the 64 bitcoin were worth approximately $4.4 million.[18]

No doubt that this development gave would-be hackers pause. It heightened doubts about whether Bitcoin transactions were truly untraceable. Moreover, it demonstrated an important lesson: unlike cash, digital currency transactions can be tracked and traced. Had Colonial Pipeline paid the hackers in cash, it's doubtful that any of the ransom funds would have been recovered.

Bad actors will always exist. At least with blockchain-backed digital assets and cryptocurrencies, there are ways to catch them.

This concludes Part I. It explained the scaffolding behind banks, payments, and the financial system. It's now time to list the benefits of making and accepting payments with digital currencies.

Chapter Summary

- Bitcoin wasn't the first type of electronic money to eliminate the need for a central authority, but it was the first to solve the double-spend problem without the role of a central authority to ensure trust and identity.

- *Digital asset* is an umbrella term that describes any intangible asset that's created and stored in digital format. This book is about digital currencies, a term that includes cryptocurrencies, digital tokens, stablecoins, and central bank digital currencies.
- As is always the case, bad actors find innovative uses for new technologies, including digital assets, to engage in criminal activity. The fleas come with the dog. However, the traceability of blockchain activities provides a better opportunity to catch criminals than when they use cash.

Part II:

The Pillars of Payments

Chapter 5

Pillar #1: The Fundamental Ability to Pay

"Anyone who has ever struggled with poverty knows how extremely expensive it is to be poor."

—JAMES BALDWIN

Ava* is a 32-year-old single mother living in the Mississippi Delta, a part of the state between the Mississippi and Yazoo Rivers. She has worked as a clerk at a local supermarket for the past four years. She currently earns $10 per hour, which is close to the state's minimum wage.

A few years ago, her then-nine-year-old daughter Mindy began experiencing digestion problems. At school one day, Mindy collapsed and needed emergency surgery. Doctors inserted a gastrostomy tube into Mindy's stomach for supplemental feeding. Fortunately, the procedure worked, and she soon recovered.

* Ava is a composite based on multiple profiles.

However, it cost Ava $3,000 out of pocket because insurance covered only 40 percent of the procedure.

An unanticipated $3,000 bill represented Ava's worst nightmare. Her FICO score was a dismal 350, and her savings were nearly nonexistent. Like roughly 60 percent of all Americans, Ava lived paycheck to paycheck. She had no friends or family to help, so she took out a single-payment title loan on her old beat-up Ford F-150. Desperate for the cash, she didn't read the fine print for the loan.

A month later, the title loan company sent her the bill. She owed the principal plus 25 percent interest: $3,750. (With these types of loans, annualized interest rates of 300 percent are sadly common.[1])

Ava couldn't make the payment. She defaulted on the loan, her car was repossessed, and she was forced to take the bus everywhere. Adding salt to the wound, her credit took yet another hit. She began relying on a combination of cash, money orders, and prepaid debit cards to pay her bills. She learned to avoid the latter because of the onerous activation, transaction, ATM withdrawal, reloading, monthly, and even inactivity fees. Industry folks refer to this as *paying to pay.*

A few years ago, Ava attempted to open a checking account at a local bank. Based on her credit history and income, the bank required her to maintain a minimum balance of $200, which wasn't possible for her. The experience further fueled her distrust of financial institutions.

Ava would love to get a job at the nearest Walmart. The company's InstaPay program lets employees access up to half their post-tax earnings before payday. Being able to spend that money even a few days early would make a big difference in her life. Without a car, however, that's not an option. It would take Ava an hour and a half to get there each way, and she can't afford childcare for Mindy. Ubers and Lyfts are too expensive for her.

Dismiss Ava as an outlier if you like, but she's not. Twenty percent of Americans can relate to her plight. Approximately 63 million Americans lack access to essential financial services and products that many take for granted.[2] And these folks are everywhere, from remote rural areas in the South to large cities in the North. Case in point: nearly 10 percent of New York City residents in 2021 lacked a checking or savings account.[3]

As we saw in Part I, private institutions and governments have spent trillions of dollars building the modern financial and banking systems. Despite their considerable efforts, a significant portion of the world's population is on the outside looking in. These folks are unable to quickly pay for basic goods and services.

Fortunately, there's a solution. Payments with digital currencies are correcting this imbalance. They enable people outside the banking system to engage in basic commerce.

Let's dive in.

The Banking Breakdown

Ava's story dispels a common myth: anyone who wants to participate in the banking and financial system easily can. As we'll see in this chapter, nothing could be further from the truth.

THE UNBANKED

I suspect that many readers of *Reimagining Payments* haven't given a lot of thought to how they make and receive payments and how much those payments cost. For example, consider depositing or cashing your paychecks. Perhaps your company has been remitting your biweekly paychecks to your savings account for the past twenty years. Your employer likely doesn't charge you a fee for the service.

Unfortunately, many of the world's population can't easily obtain checking or savings accounts or easily make cashless payments. We call these individuals the *unbanked*.

In 2021, the Federal Deposit Insurance Corporation released the results of its biannual *National Survey of Unbanked and Underbanked Households*.[4] The FDIC found that low-income, less-educated households are far more likely to be unbanked than affluent, educated households. The same holds true for Black and Hispanic families compared to white families. Finally, households led by people with disabilities and single mothers skew heavily toward the unbanked.

That same FDIC survey revealed that an estimated 4.5 percent of US households were unbanked: roughly 5.9 million. It's a sizeable number but a pittance compared to the rest of the world.

By late 2022, the world's population topped 8 billion.[5] Globally, of approximately 6 billion adults, about 1.4 billion are unbanked, which means that 23.3 percent, or nearly one in four adults, are unbanked.[6]

Across the globe, you're likely to find plenty of people like our fictional friend Ava, but the unbanked aren't evenly distributed.

For example, in Nordic countries such as Sweden, Finland, and Denmark, it's rare for an adult to lack a bank account.[7] In other sovereign nations, the opposite is true. Figure 5.1 shows the nations with the highest percentage of unbanked adults.

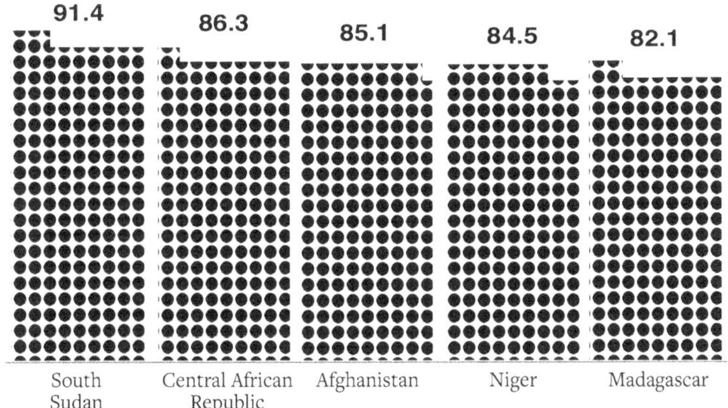

Figure 5.1: Percentage of Unbanked Adults
Source: The Global Findex Database I https://tinyurl.com/32s3fd6x

Regardless of where they live, the unbanked still need to engage in the financial activities you and I might take for granted. In other words, they must meet their financial needs *outside the current system.*

Alternative Financial Services

We've now arrived at alternative financial services. AFSs provide expensive last-resort alternatives for people who willingly or unwillingly operate outside the financial system. Examples include Clearviewloans, Bad Credit Loans, and Speedy Cash. As the name connotes, AFSs include the following bank-like services:

- **Transaction-based services:** These include cashing checks, issuing money orders, and remitting funds internationally.

- **Credit-based services:** These include payday, refund-anticipation, pawn shop, and auto title loans.[8]

Despite their prohibitively expensive nature, desperate folks often use AFSs anyway. The word *usury* comes to mind. (Ava's 300-percent annual interest rate mentioned at the start of this chapter is alarming but, sadly, not uncommon.)

ProPublica and The Current published an in-depth investigation on the business practices of TitleMax, a large private lender based in the southern US. The company lends money in exchange for collateral, such as the title to the vehicle the customer drove to the store. This is referred to as *title pawns*. Former store managers at some of TitleMax's 900-plus locations "were trained to keep customers unaware of the true costs of their title pawns. When they were more transparent, they faced repercussions."[9]

TitleMax has been on the US government's radar for years. The Consumer Financial Protection Bureau fined its parent company $9 million in 2016 for violating federal laws by engaging in predatory lending practices in a few states.

There's no shortage of AFSs in the US, and you're likely to find a Money Mart, Check Into Cash, National Cash Advance, and Cash America in impoverished neighborhoods. These businesses wouldn't exist if they weren't profitable—and they wouldn't be profitable if people didn't need to use them.

For the unbanked, even the act of cashing a government-issued check poses formidable obstacles. In March 2020, the US government distributed $271 billion in stimulus aid to American citizens as part of the CARES Act. Millions of unbanked recipients relied on expensive AFSs to cash these checks. These folks cumulatively paid an estimated $66 million in fees.[10]

These onerous fees often force the poor to reluctantly draw against already-low balances. Others incur overdraft fees. In 2021, the Consumer Financial Protection Bureau reported that:

> Banks continue to rely heavily on overdraft and non-sufficient funds (NSF) revenue, which reached an estimated $15.47 billion in 2019 .... Three banks—JPMorgan Chase, Wells Fargo, and Bank of America—brought in 44% of the total reported that year by banks with assets over $1 billion. The CFPB also found that while small institutions with overdraft programs charged lower fees on average, consumer outcomes were similar to those found at larger banks. The research also notes that, despite a drop in fees collected, many of the fee harvesting practices persisted during the COVID-19 pandemic.[11]

Understanding the Unbanked

Given AFS's considerable fees, the natural question is this: why do they have customers? There are two main reasons. First, some people are unaware of cheaper alternatives for check cashing and lending services. Second, many folks simply don't have a choice because they're unbanked or they live in an area without a local bank. In such places, you're likely to find plenty of AFSs.

As for why people visit AFSs instead of banks and credit unions, Figure 5.2 provides some insights.

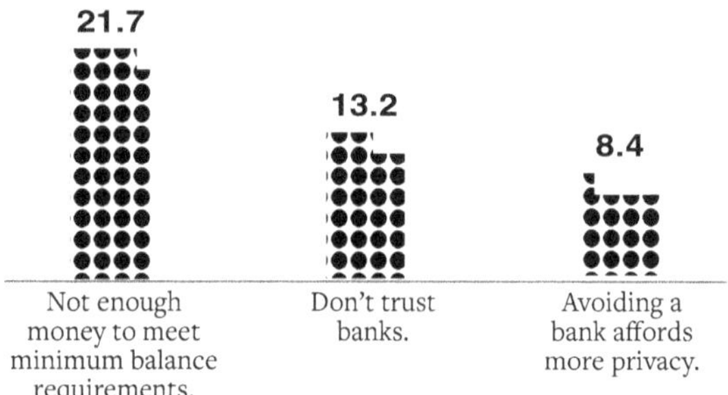

Figure 5.2: Most Cited Reasons for Why People Don't Have a Bank Account

THE FULLY BANKED

Now let's go to the opposite end of the spectrum. If you're reading this book and you're a US citizen, the odds are high that you qualify as *fully banked*. That is, you meet the following two conditions:

- You hold a checking or savings account with an FDIC-insured institution like Wells Fargo, Chase, or Bank of America.
- You haven't used an AFS in the past 12 months.

THE UNDERBANKED

Sitting in between the fully banked and the unbanked are the underbanked. In the US, the underbanked have a checking or savings account with an FDIC-insured institution but still regularly use AFSs. In 2021, the FDIC found that 14.1 percent of US households (approximately 18.7 million) used at least one of the following nonbank transaction or credit products or services that are disproportionately used by unbanked households to meet their transaction and credit needs: money orders, check cashing, international remittances (i.e., nonbank transactions), or rent-to-own services or payday, pawn shop, tax refund anticipation, or auto title loans (i.e., nonbank credit).[12]

Figure 5.3 shows these three groups as part of the same continuum.

Figure 5.3: The Banking Continuum

Financial Inclusion and Why It Matters

Regardless of where they live, the unbanked can't participate in the banking and financial system. The underbanked are more likely to *willingly* forgo transacting with banks and credit unions. Or, they have limited access to their bank accounts to facilitate basic financial transactions and to obtain credit. Put differently, the current financial system *excludes* a significant percentage of the world's adult population.

DEFINITION

We've now landed on a critical concept and massive limitation of the current banking system: *financial inclusion*. The World Bank defines it as follows:

> individuals and businesses have access to useful and affordable financial products and services that meet their needs—transactions, payments, savings, credit, and insurance—delivered in a responsible and sustainable way.[13]

Broadly speaking, inclusion here represents the ability of previously excluded and underserved populations to access essential financial products and services.[14] Examples include:

- Interest-bearing savings accounts.
- Checking accounts and the ability to write checks.
- Credit cards.
- Insurance.
- Investment products.
- The ability to borrow money from financial institutions.

Increasing financial inclusion is a worthy goal with important economic, political, and social implications. For example, consider the Group of Twenty. The intergovernmental forum consists of nineteen countries and the European Union. The G20 attempts to address global economic issues, including international financial stability, climate change, and sustainable development.

Achieving universal financial inclusion is a Sisyphean task that has vexed policymakers for decades. The G20 lists it as one of its explicit objectives, but how can society achieve it?

There isn't a single or simple answer to this question. If we're to achieve universal financial inclusion, technology will play a prominent role.

TECHNOLOGY

In 2011, the esteemed venture capitalist and browser pioneer Marc Andreessen famously said, "Software is eating the world."[15] He wasn't discussing financial inclusion, but his words were eerily prescient. Streaming services replaced video stores like Blockbuster. Uber and Lyft decimated local taxi services. The Blackberry disappeared. You get my drift.

As we'll see throughout Part II, powerful digital tools are providing AFSs to hundreds of millions of people who are unbanked, underbanked, and banked. Technology, software, and digital currencies are already opening up the financial system to those on the outside looking in.

By the same token, one can't fully understand—much less address—the subject of financial inclusion today without considering the profound impact of technology.

To its credit, the G20 has long understood this point. From its 2016 *High-Level Principles for Digital Financial Inclusion:*

> Leveraging the opportunities that technology offers to reduce costs, expand scale, and deepen the reach of financial services will be critical to achieving universal financial inclusion.[16]

Let's delve into a few examples of how payment technology is leveling the playing field.

How Technology and Digital Currencies Are Increasing Financial Inclusion

Think about your first cellphone. Depending on your age, it may have been a Nokia 6110, a Motorola StarTAC, or a Blackberry.* It's no overstatement to say that even early entrants were enormously helpful. They altered many aspects of our lives—with many more to come.

The mobile phones of the late 1980s and 1990s were indeed powerful, but they were primitive when compared to the iPhone, Samsung Galaxy, Google Pixel 7 Pro, and other modern counterparts. Today we use these mini-computers to do so much more than text a friend at a loud, crowded bar or snap a quick photo or

* Take a trip down memory lane at https://tinyurl.com/repay-phone.

video.* These early, relatively banal uses of smartphones pale in comparison to the transformative impact that simple cell phones have had in certain underdeveloped countries.†

M-PESA

For years, Vodafone's Kenyan associate, Safaricom, has been Africa's largest mobile phone operator. In 2007, it launched M-Pesa—a phone-based system that allows people to transfer money without the friction of alternative financial services. (In case you're curious about its name, *M* stands for *mobile*, and *pesa* is Swahili for *money*.)

The mobile money service has more than 604,000 active agents operating across the Democratic Republic of Congo (DRC), Egypt, Ghana, Kenya, Lesotho, Mozambique, and Tanzania.[17] Not surprisingly, there's also an M-Pesa app for smartphones.

Think of M-Pesa as an alternative way for Africans to access essential financial services easily and affordably, including:

- Depositing and withdrawing local currency.
- Transferring funds to other users.
- Receiving transfers from other users.
- Paying bills.
- Storing currencies in virtual accounts, called *M-Shwaris.*
- Transferring currencies between the service and, depending on the country, a bank account.

---

* Check out https://tinyurl.com/repay-iphone for a list of things that smartphones have replaced.

† Underdeveloped countries are those with less stable economies, less democratic political regimes, greater poverty, malnutrition, and poorer public health and education systems.

Innovation and the Adjacent Possible

Before continuing, it's essential to point out that M-Pesa isn't a digital currency. I'm referencing it here because of its success as an early mechanism to let the unbanked make payments in new ways. The same thing will happen with digital currencies because the opportunities are vast.

The advent of cell phones opened up entirely new possibilities. The same goes for communications technologies. As Steven B. Johnson describes in his 2010 bestseller *Where Good Ideas Come From*, these types of innovations expand the adjacent possible.

As we'll see throughout this book, blockchain technology and digital currencies similarly allow for new ideas around payments.

M-Pesa in Action: A Simple Example

Since its inception, M-Pesa has grown at a blistering pace. Today an astonishing 96 percent of Kenyan households use it.[18] Consider the following example.

John Mwangi is a 52-year-old fisherman who lives in a remote village outside of Nairobi, Kenya's capital and its largest city. The nearest local banks are thirty miles away. As a result, he's never held a bank account.

At the end of a day fishing, Mwangi takes his catches to Betty Odhiambo, an amiable local market owner, for purchase. The price for today's haul is 800 shillings.

To complete the transaction, neither party needs to visit a bank. Odhiambo doesn't have to carry a bundle of shillings to pay suppliers like Mwangi. Instead, Odhiambo simply accesses the funds she added to her M-Pesa account at one of the country's agent outlets.

Odhiambo pulls up Mwangi's phone number, enters her personal identification number (PIN), and remits the 800 shillings. Within seconds, both parties receive SMS notifications* on their cellphones confirming the following information:

- The amount of the transaction.
- The name of the other party involved in the transaction.
- The new balances in each party's account.

Results

It's hard to overstate the impact of M-Pesa on the Kenyan economy. The unbanked can now do simple yet essential things that were previously far more difficult and expensive. As one example of how M-Pesa has revolutionized the way an entire country pays for and receives goods and services, consider Figure 5.4.

These formidable numbers mask the broader economic impact that M-Pesa has had on Kenyan society. Researchers in a 2014 paper found that its increased use:

- Lowers the propensity of people to use informal savings mechanisms.
- Raises the probability of their being banked.
- Promotes banking and commerce.
- Allows micro businesses to flourish.[19]

---

* The acronym SMS stands for short message service, although we commonly refer to them as *text messages* or *texts* these days.

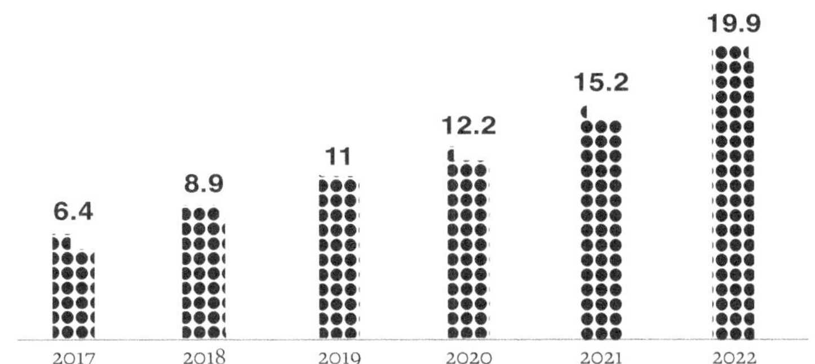

Figure 5.4: M-Pesa Transaction Volume From 2017 to 2022 (In Billions)
Source: Statista

It gets even better. A 2016 study found that access to M-Pesa increased per-capita consumption levels. It also lifted 194,000 Kenyan households out of poverty. That's 2 percent of the country's population.[20]

Importantly, M-Pesa derives serious revenue from these activities. Based on the M-Pesa business model, my thesis is that there's a significant revenue-generating opportunity for digital currency payments.

UKRAINE

Smartphones and payment apps like Venmo and Zelle have already reduced the number of unbanked people worldwide. As we'll see next, we're just scratching the surface of what digital currencies can do. Confronted with a global crisis, payment pioneers quickly found innovative ways to help a country under siege securely remit funds to its citizens and military.

Background

February 24, 2022, when Russian forces invaded Ukraine without provocation, is a day that will live in infamy. Peace talks have failed. The US, Great Britain, Germany, and other Ukrainian allies have provided invaluable financial and military assistance.

Military aggression is old hat and, unfortunately, even seems to be making a comeback these days. The way that many concerned citizens across the globe responded to the Ukraine invasion, however, has been unprecedented. For our purposes, it represents a critical case study.

Everyday people wanted to help Ukraine however they could. As expected, the Red Cross and Ukrainian Red Cross Society accepted donations. Others started grassroots fundraising projects. As it turned out, digital currencies proved remarkably popular and effective for these purposes.

Results

Days after the invasion, Ukraine's official X (formerly Twitter) account began sharing two addresses linked to its digital currency wallets.[21] Four days after the invasion, people worldwide had sent 153 Bitcoin (worth $6.29 million) and 2,230 ETH (worth $6.27 million) to the addresses.[22] Interestingly, anyone can view the current state of Ukraine's crypto donations in real time.[23]

Ukraine is using these funds to support its military and purchase supplies.[24] The country "bought a total of 416,900 field rations through crypto donations."[25] For security reasons, Ukraine's government has wisely chosen not to be more specific about how it dispersed funds raised.

If digital currencies can help during a war, why not during peace?

AIDING THE AFGHANS

On August 31, 2021, the Biden Administration announced that the US was pulling out of Afghanistan, ending twenty years of conflict. The Taliban quickly assumed control, and the US government responded by imposing sanctions on Afghanistan.

The sanctions meant that the Taliban could not access roughly $10 billion in reserves held in the Afghan central bank. That money was effectively frozen; the trickle-down effects of the US sanctions were immediate and profound. A full-blown humanitarian crisis ensued.

As a result of the US sanctions, Afghanistan's economy has teetered on the verge of collapse. The Taliban has severely restricted the ability of its citizens to withdraw funds from their bank accounts.

The Importance of Cash

It's difficult to overstate the importance of cash in Afghanistan. (Its official currency is the *afghani.*) It's a largely cash-based society, and businesses often accept no other form of payment. In 2017, fewer than one in four male adults owned a valid bank account. The corresponding number for female adults is 7 percent.[26]

Afghans today face a shortage of paper money, and millions of people can't afford to buy food or medicine. The situation is dire. The World Food Programme estimated in late 2021 that 98 percent of Afghans lacked sufficient sustenance.[27] Facing starvation, some Afghans sold their kidneys.[28] In December 2022, the UN Security Council attempted to address the crisis by:

> Adopt[ing] a broad exception to the Afghanistan financial sanctions, covering the provision, payment, and processing

> of funds and assets necessary for humanitarian action and for activities to meet basic human needs.[29]

No one knew whether the carve-out would ultimately work and how long it would take. Afghani Fereshteh Forough wasn't about to wait to find out.

Thinking Outside the Box

Along with her family, Forough fled Afghanistan after the Soviet invasion in the early 1980s. Today she's a social activist living in the northeastern US. In January 2015, she founded Code to Inspire, the first coding school for girls, located in Herāt.* Forough has taught more than 150 girls how to code.

Forough saw what was happening in her home country and felt compelled to help—even from 6,000 miles away. She first tried remitting emergency payments to her former and current students at Code to Inspire. Against a backdrop of US sanctions, however, US financial institutions initially refused to let the transactions through. Eventually, some of her transfers succeeded, but it was a moot point: recipients couldn't withdraw cash at their local banks.

Frustrated, Forough got creative to support needy students at Code to Inspire. She started sending them digital currencies.[30] As she told Insider in January 2022:

> We found that actually there are a bunch of local money exchanges in the financial district of Herāt that are accepting crypto and they can cash it out for you in either afghanis or dollars.

---

* Herāt is the third-largest city of Afghanistan.

It's an inspiring story. Since September 2021, Forough and her team have been sending the cash equivalent of $200 per family per month in digital currency.[31]

If digital currencies can help the most marginalized people in a society during a time of need, why not help entire populations whether in need or not?

TYING IT ALL TOGETHER

As we've seen in this chapter, payments using digital currencies can increase financial inclusion and improve day-to-day life for marginalized individuals and companies that are excluded from traditional financial services and banking institutions. Digital currencies represent an exciting and opportunistic way for businesses to expand their customer base and bring payments into the next century while turning a profit.

Chapter Summary

- Globally, billions of people are excluded from traditional financial services offered by banks and credit unions.
- Their options for cashing checks, using money orders, remitting funds internationally, and obtaining credit are primarily costly (and sometimes deceptive and predatory) alternative financial services.
- Digital currency payments can improve financial inclusion. They can expand access, ownership, and usage of a broad range of financial services.
- Digital currency payments represent an opportunity to increase the customer base and market share while supporting financial inclusion.

Chapter 6

Pillar #2: Security and Privacy

"There are two types of companies: those that have been hacked, and those who don't know they have been hacked."

—JOHN CHAMBERS

Unless your job involves information security, the date March 10, 2017, most likely doesn't jump out at you. There's a good chance, though, that you've had to deal with the consequences of that fateful day—or will at some point. On that otherwise uneventful Friday, hackers penetrated the networks of the credit bureau Equifax.

First, as is often the case with security incidents, the company didn't immediately realize that bad actors had compromised its systems. And Equifax only revealed the attack on September 7, 2017, nearly six months after it took place and after weeks of internal forensic analysis confirmed it.

Next, despite previous known breaches, Equifax failed to adequately secure its internal systems and processes.[1] Finally, the hackers intelligently covered their tracks. As Josh Fruhlinger wrote in a telling postmortem:

> Equifax's attackers encrypted the data they were moving in order to make it harder for admins to spot; like many large enterprises, Equifax had tools that decrypted, analyzed, and then re-encrypted internal network traffic, specifically to sniff out data exfiltration events like this.[2]

It's a sad state of affairs. Thanks to decades of hacks and other security incidents, many of us have developed immunity to them.

For a few reasons, however, the Equifax attack was unique. In this lengthy chapter, I'll explain why—and how digital currencies offer a fundamentally more private and secure alternative to traditional payment methods.

An Especially Horrendous Hack

The scope of the Equifax attack was unprecedented. Analysis revealed that bad actors accessed the sensitive information of a mind-boggling 147 million Americans. To this day, few hacks rival its size. (In case you're curious, Yahoo! holds the record as of this writing, sporting a breach of 3 billion accounts.[3]) Appalling, to be sure, but it gets even worse.

By way of background, breaches solely involving usernames and passwords are certainly dangerous. (If you're one of the few who has never been a victim of online fraud, just use your imagination. You're playing Russian roulette if you haven't enabled two-factor authentication on your most sensitive accounts.) Once the compromised organization sends its users emails or letters, users generally change their login information.

Resetting our passwords typically involves jumping through a few hoops, but the jumping is worth it: the process ultimately minimizes the damage and prevents further disruption to our lives.

Put differently, our usernames and passwords are *ephemeral.* There's nothing precious or immutable about them. Our names, dates of birth, and social security numbers, however, are entirely different and much stickier. The Equifax hackers grabbed this *permanent* information, along with customers' home addresses and credit card numbers.

When all was said and done, the scandal—and its backlash—forced several of its executives to depart, including CEO Richard F. Smith.[4] Equifax's reputation also took a hit. As expected, plaintiffs' lawyers and regulators quickly pounced. To settle the class-action lawsuit, Equifax wrote a $575 million check and:

> ... agreed to a global settlement with the Federal Trade Commission, the Consumer Financial Protection Bureau, and 50 US states and territories. The settlement includes up to $425 million to help people affected by the data breach.[5]

Equifax also tossed in free credit monitoring for its victims,[6] but that concession didn't quell the outrage. As Farhad Manjoo wrote in the *New York Times*:

> Equifax, you had one job. Your only purpose as a corporation, the reason you were created and remain a going concern, is to collect and maintain people's most private financial data.[7]

Manjoo is spot-on, but it's essential to note Equifax is hardly the only company that failed to protect its customers' valuable data. Look at this smattering of other examples:

- In 2013, Chase reported that hackers accessed data of 465,000 prepaid cards used for payroll and government benefits.[8]
- In 2019, former Amazon engineer Paige Thompson downloaded data from more than 100 million Capital One users. The breach included 120,000 Social Security numbers and nearly 80,000 bank account numbers. US Attorney Nick Brown claimed that Thompson "did more than $250 million in damage to companies and individuals."[9]
- In May 2019, as a result of a website design error, First American Financial Corporation exposed nearly 900 million "financial and personal records linked to real estate transactions."[10]
- In 2019, Marriott admitted that its Starwood hotel unit didn't encrypt passport numbers for roughly five million guests and the numbers were lost in an attack. Experts believe the attack was carried out by Chinese intelligence agencies.[11]

And let's not forget the money spent to contain the damage. In 2021, the Treasury Department's Financial Crimes Enforcement Network revealed that US banks and financial institutions processed roughly $1.2 billion in ransomware payments that year.[12]

I could keep going, but you get the point: hackers have long targeted large banks, credit card companies, credit bureaus, and other companies that retain customer data.*

Privacy and the Status Quo

Interestingly, cyberattacks don't affect all US citizens equally and to the same degree. When breaches occur, the unbanked and underbanked—discussed in Chapter 5—typically don't suffer to the extent that fully banked Americans do because they haven't entrusted an institution with their personally identifiable information (PII). It's difficult to steal financial information from folks who can't establish bank accounts, obtain credit cards, and the like.

The US Department of Labor defines PII as:

> Any representation of information that permits the identity of an individual to whom the information applies to be reasonably inferred by either direct or indirect means.[13]

PII certainly existed before the arrival of the commercial internet in 1994. In hindsight, however, few people truly appreciated the effect that ecommerce would have on the privacy and security of our personal and professional lives. The following sidebar represents just one example of how these three forces collided.

* See https://tinyurl.com/repay-hacks for an updated list of the biggest known breaches.

Facebook's Beacon Debacle

For its first three-plus years, the world's largest social network generated zero revenue. Facebook—now Meta—charged its users nothing, and CEO and founder Mark Zuckerberg refused to run ads. Industry veterans knew that the clock was ticking. It was only a matter of time before something had to give.

On November 6, 2007, the company introduced Facebook ads and a component called Beacon, an "ad system for businesses to connect with users and target advertising to the exact audiences they want."[14]

Facebook was trying to revolutionize advertising and cash in on its massive user base. Beacon let users broadcast purchases made on third-party websites for all their Facebook friends to see.

What could go wrong? In a word, lots.

Facebook enabled Beacon by default. That is, users had to opt *out* of the feature. As a result, Beacon-affiliated sites (such as Fandango and Overstock.com) enabled the "friends" of Facebook users to automatically receive a notification when a user made a purchase. This notification happened before the purchaser had a chance to approve it.

User complaints abounded. As Betsy Schiffman wrote in *Wired*, "Facebook apparently never considered that sometimes people want to keep their shopping habits to themselves."[15]

Zuckerberg apologized and promised to do better.

OUR EVOLVING ATTITUDES TOWARD PRIVACY

Still, Zuckerberg and Meta are hardly alone in amassing and using vast troves of data to fuel their businesses. Data brokers like Acxiom make their money aggregating and selling data. Google's trove of data rivals Meta's, and both entities have had breaches. Even Netflix purchases third-party data and metadata to fine-tune its algorithm to provide its subscribers with better streaming recommendations.[16]

Returning to Zuckerberg momentarily, consider what he told a *Time* reporter in 2010: "What people want isn't complete privacy; they want control."[17]

It was an interesting thesis, but did it reflect society's widespread beliefs back then? And what about now? Do we want to use our apps, surf the web, make payments, and perform other activities online without companies tracking our behavior?

A plethora of surveys have asked respondents about their online privacy preferences. What would users do when provided with a real choice about sharing their online activities with third parties?

A case study of Apple's release of version iOS 14.5 provides an answer to Zuckerberg's question from 2010 in earnest.

On April 26, 2021, Apple released version iOS 14.5, which required developers to obtain users' permission to track their online behavior. After upgrading to the new version of iOS and launching their apps, iPhone users started seeing the prompt displayed in Figure 6.1.

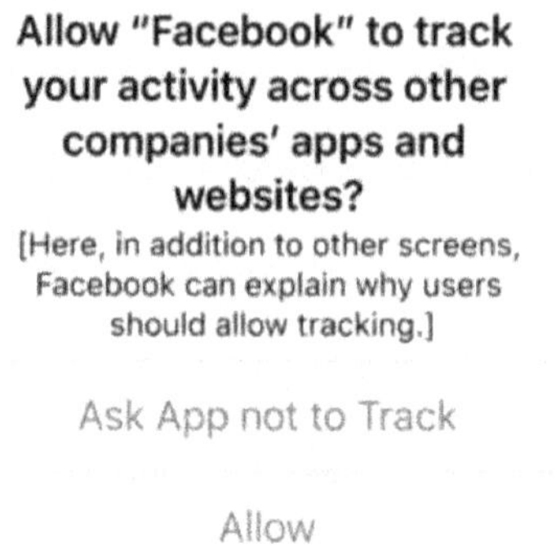

Allow "Facebook" to track your activity across other companies' apps and websites?

[Here, in addition to other screens, Facebook can explain why users should allow tracking.]

Ask App not to Track

Allow

Figure 6.1: iOS 14.5 Tracking Prompt
Source: iOS/Apple

How would people respond? Did they really want control? Were Zuckerberg's 2010 comments prescient?

The results were telling. A remarkable 96 percent of US users opted out of app tracking.[18] For example, local restaurants that used to rely upon location data to offer specials and coupons no longer knew the locations of their prospective customers. Without users' consent, Facebook ads ceased to be (as) effective.

Put differently, by allowing users to opt out of tracking, Apple had effectively kneecapped its rival's ad business in one fell swoop. Meta's 2022 revenue plunged by $10 billion.[19] At the time, 90 percent of its revenue stemmed from targeted ads. Figure 6.2 shows the dip.

The nearly unanimous response reflected a new reality: we're increasingly rethinking our beliefs about online privacy. In April 2020, Pew reported that half of Americans "decided not to use a product or service because of privacy concerns."[20]

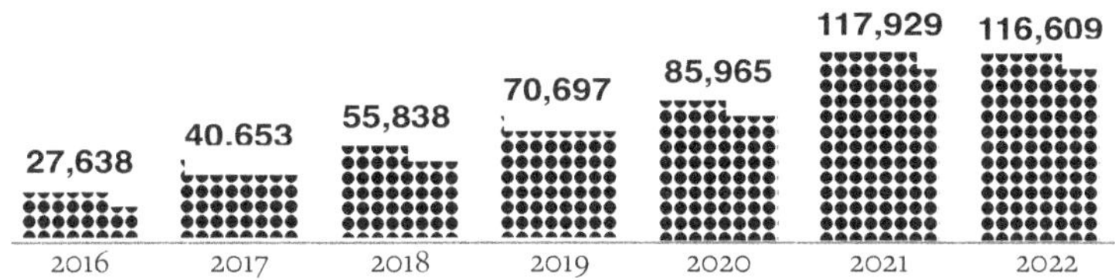

Figure 6.2: Annual Revenue Generated by Meta Platforms From 2016 to 2022 (In Millions)
Source: Statista

Additional research from Pew in November 2019 reveals similar findings:

- More than four in five Americans say that the potential risks of data collection outweigh the benefits of companies' free services.
- Two-thirds say the same about government data collection.
- Seventy-nine percent of Americans report being concerned about how companies use their data.
- Sixty-four percent say the same about the government.[21]

The people have spoken. People crave some degree of privacy after all. In other words, we're rejecting *surveillance capitalism*, to borrow from the title of Shoshana Zuboff's bestselling 2019 book.

The need to reclaim online privacy is understandable. Some argue that when it comes to making digital payments, absolute privacy is needed as an alternative to transacting in government-issued cash. Although cash transactions are completely private, users face practical obstacles when trying to pay for goods and services using cash. For example, consider large-value transactions. It would be difficult—though not impossible—to purchase a car for $60,000 in physical currency.

Plenty of folks believe that transactions in Bitcoin are *anonymous*. Nothing could be further from the truth. Instead, they're *pseudonymous*. In other words, transaction addresses mask users' identities. But with the proper tools and techniques, you can trace the transaction back to the owner of the transaction address.

Moreover, there are potential dangers to complete payment privacy. Let's explore why.

DIGITAL CURRENCY PAYMENT PRIVACY

As a general proposition, financial institutions don't want to facilitate malfeasance, much less criminal activity. They also don't want to circumvent the applicable laws in the countries and states in which they operate. We discuss relevant laws and regulations in Chapter 9.

In the US, the Financial Crimes Enforcement Network (FinCEN) regulates more than 100,000 financial institutions, including banks, credit unions, money services businesses (MSBs), insurance companies, securities brokers, and casinos. The government has deemed these organizations at risk of being used by criminals to support illicit enterprises such as drug cartels, mortgage fraud rings, and terrorist finance networks.

FinCEN administers the Bank Secrecy Act (BSA). Passed in 1970, the law requires US financial institutions to assist US government agencies to detect and prevent money laundering. The BSA is sometimes referred to as an *anti-money laundering* (AML) law or jointly as "BSA/AML."

The BSA requires, among other things, financial institutions to appoint a chief compliance officer, implement certain policies and procedures, make filings with the government about certain types of transactions, and keep certain records. Among others, these records include:

- Cash purchases of negotiable instruments that guarantee the payment of a specific amount of money.
- Reports of cash transactions exceeding a daily aggregate amount of $10,000.
- Reports of suspicious activity that might signify money laundering, tax evasion, or other criminal activities.

Similarly, the International Monetary Fund recognizes the importance of guarding against money laundering and terrorism financing activities by collecting data to thwart these activities. From its website:

> An effective anti-money laundering/counter-financing of terrorism framework must therefore address both risk issues: it must prevent, detect, and punish illegal funds entering the financial system and the funding of terrorist individuals, organizations and/or activities. Also, AML and CFT* strategies converge; they aim at attacking the criminal or terrorist organization through its financial activities, and use the financial trail to identify the various components of the criminal or terrorist network. This implies to put in place mechanisms to read all financial transactions, and to detect suspicious financial transfers.[22]

Accurate account and transaction data provide a trail of financial transactions—digital breadcrumbs that authorities use to catch the bad guys.

---

* Short for *counter-financing of terrorism*.

For US transactions, financial institutions must generally collect the name, date of birth, address, and a government identification number (such as a driver's license or passport number) of customers.

As a result, payments using anything other than cash, such as debit cards, credit cards, bank account transfers, smartphones, and apps, lack privacy *by design*.

THE POPULARITY OF DIGITAL PAYMENTS

Paying bills with physical money certainly provides anonymity from prying digital eyes. Fine, but delivering cash to brick-and-mortar stores is time-consuming and potentially dangerous. No one would call obtaining a money order or a cashier's check *a seamless process*. Neither is hauling $60,000 in cash to a car dealer to purchase your vehicle of choice.

Put differently, despite increased privacy concerns, most of us don't want to forgo the convenience of electronic payments, and the data bears that out.

McKinsey's 2021 *Digital Payments Consumer Survey* found that more than 80 percent of Americans used some form of digital payment in 2021.[23] Popular methods included:

- Browser-based purchases.
- In-app purchases.
- In-store checkouts via smartphones.
- Quick-response (QR) codes.
- Person-to-person (P2P) payments.

All people should be able to purchase legal goods and services securely and privately without forgoing convenience in a manner that complies with all relevant regulations.

A Better Mousetrap: Unpacking Digital Currency Security

Privacy and security are inextricably linked. It's hard to envision the former without the latter. It's time to explain some of the most important security concepts behind digital currencies and payments made with them.

Author's note: this part of the chapter provides technical information relating to digital currency payments. As such, feel free to skim these sections if desired.

CUSTODIANS

Custody is a fascinating word with—no shocker here—Latin roots.* Merriam-Webster defines it as "immediate charge and control (as over a ward or a suspect) exercised by a person or an authority."[24] At some point you've heard a news anchor say, "Police captured the suspect, and he's now in custody." In a different vein, odds are that you know a divorced couple that share custody of their kid(s), or maybe one parent retains primary custody.

Apart from criminals and children, custody applies to the financial services area as well. In this context, a custodian—or *custodial service*—safeguards its clients' cash, securities, gold bars, and other assets. Traditional custodians have operated since the 1960s. They serve as key pillars of our financial system.

There are custodians in the digital currency arena as well, such as Coinbase Custody, Gemini, Fidelity, and Etana Custody.

---

* It stems from the Latin word *custōdia*, meaning "a keeping, watch, guard, prison."

The reason isn't complicated: some people prefer that a trusted third party protects their digital assets. In this way, crypto custodians serve a critical function: they take control—or *custody*—of their clients' digital currency holdings.

WALLETS

Say that it's 1970. You take your significant other out to dinner. When the check arrives, you decide to pay. You prefer to use cash, but the amount of the bill exceeds your cash on hand. Rather than risk embarrassment, you give the waiter your credit card.

Fiat currencies and debit and credit cards share one key attribute: they're physical. You can hold them in your hands. You can store each of these payment methods in a tried-and-true physical storage container: a wallet.

At a high level, wallets serve a similar purpose. Think of them as storage locations that allow people to access their digital currencies freely, easily, and securely.

Wallets don't physically hold digital currencies. Instead, they store the private keys (explained below) required to transact with digital currencies and provide digital signatures that authorize each transaction.

Two core attributes are inherent in all of them. The first involves connectivity, of which there are two dimensions:

- **Hot:** Online; the private keys are accessible to software that has an internet connection. While easy and convenient to use, they're vulnerable to hacking and online attacks.
- **Cold:** Offline; the private keys aren't accessible to software that has an internet connection. They're less

convenient to use, but they're also less susceptible to hacks and online attacks.

The second attribute centers upon hosting of the wallet:

- **Self-Custody:** The user controls the private keys; they're more secure because funds are stored offline and are less prone to hacks and attacks. If the user loses their private keys, they lose access to their funds.
- **Third-Party Hosted:** A third-party controls the private keys. These wallets require the user to trust that the third party will keep the private keys secure. If the user loses their private keys, the third-party can recover them.

Table 6.1 fuses these two attributes and dimensions. In so doing, it displays examples of the four main types of wallets.

Type of Wallet	Self-Custody	Third-Party Custody
Hot	Atomic Wallet, Metamask	Bitpanda, Coinbase
Cold	Trezor, Ledger, Ellipal	BitGo, Fireblocks, Copper

Table 6.1: Examples of Wallets

We'll return to how wallets enable secure payments with digital currencies shortly. For now, we need to take a step back.

CRYPTOGRAPHY REVISITED

Chapter 4 touched briefly on cryptography in the context of Bitcoin and blockchain technology. For the purposes of this book, let's focus on the main aspect of the subject: keys. A key is a long string of random characters.

Think of a public key like a bank account number that can be shared with others to receive funds. On the other hand, think of a private key like the password or PIN for the bank account. It provides unfettered access to the account. For that reason, private keys should be kept confidential. In asymmetric-key cryptography, every public key is paired with one corresponding private key. Together, they're used to encrypt and decrypt data for transactions.[25]

Multisignature or "multisig" wallets bear a quick mention here. Multisigs require multiple signatures from a set of predetermined addresses to transact. If the minimum required number of signatures isn't provided, the transaction can't go through. Think of it like a safe with unique keys that must be used together to open it. As such, they add a layer of security.

Moving From Theory to Practice

If Part I manifested nothing else, it's the sheer complexity of the modern payment system. Banks, credit card companies, governments, and other financial institutions have created and use payment rails that, although imperfect, allow people to conduct billions of transactions per day. Can someone say *entrenched interests*? Yet, payments with digital currencies have arrived. Here are some exciting projects.

PROJECT BAKONG

In October 2020, the National Bank of Cambodia (NBC) launched Project Bakong, a nationwide, blockchain-based payment system that represented one of the world's first central bank digital currencies (CBDCs). NBC launched the product to strengthen the national currency, the Khmer riel, and reduce dependence on the US dollar. Almost 200,000 people use its digital wallet, and almost six million people have benefited through the use of connected banking apps.

NBC runs the crux of the system along with Japanese technology company Soramitsu using the open source Hyperledger Iroha blockchain framework. The flexibility of using a blockchain for digital asset management allows NBC and Soramitsu to implement fiat-backed digital representations of the Khmer riel and US dollar that are accessible for wholesale interbank transactions and everyday retail payments.

It also allows financial institutions to run payment gateways that permit users to connect to the Project Bakong system via mobile apps. (For a refresher on gateways, see the "What's a Payment Gateway" sidebar in Chapter 2.) Blockchain's inherent security features and its decentralized nature allow Project Bakong to mitigate the risks of fraud, hardware failure, tampering, and, most importantly, cyberattacks.

For example, blockchain technology allows Project Bakong to store all transactions as a chronological chain of records and replicates this data across multiple servers. Interestingly, Project Bakong is geographically distributed or decentralized. It collects and stores all transactions but no PII. Instead, the financial institutions that join the system retain the PII. In turn, these institutions only receive transactions for which their users are counterparties; they lack access to all transactional data in the system by design.

Project Bakong intentionally creates silos to separate its customers' PII from the PII of noncustomers of participating financial institutions. In so doing, Project Bakong preserves the privacy of its users while concurrently allowing NBC to view all transactional activity. This data helps the government fine-tune its monetary policy.

The US has considered developing its own CBDC project. Count Australia, Canada, the United Kingdom, China, Japan, the United Arab Emirates, and Singapore among the others doing the same.[26]

PROJECT HAMILTON

In the summer of 2020, the Boston Federal Reserve began collaborating with the Massachusetts Institute of Technology on an exploratory research project named after the Founding Father we met in Chapter 3. The goal was simple yet ambitious: to simulate the millions of transactions people would make if the US saw widespread CBDC adoption.

In February 2022, they released the results of Project Hamilton, Phase 1. From the report's Executive Summary:

> The Federal Reserve Bank of Boston (Boston Fed) and the Massachusetts Institute of Technology's Digital Currency Initiative (MIT DCI) are collaborating on exploratory research known as Project Hamilton, a multiyear research project to explore the CBDC design space and gain a hands-on understanding of a CBDC's technical challenges and opportunities. Our primary goal was to design a core transaction processor that meets the robust speed, throughput, and fault tolerance requirements of a large retail payment system. Our secondary goal was to create a flexible platform for collaboration, data gathering, comparison with multiple architectures, and other future research. With this intent, we are releasing all software from our research publicly under the MIT open-source license.[27]

There's a good deal to unpack in the report, and I encourage you to read it yourself. For now, I'll highlight a few of Project Hamilton's baseline requirements and encouraging results.

Speed and Time to Finality (TTF)

Thanks to smartphones, social media, and ubiquitous tech, people are fundamentally impatient. Attention spans are plummeting.[28] Hoping that consumers will patiently wait for their payments to clear is optimistic at best. Double that when they have other payment options. (Chapter 7 returns to this subject.)

Speed matters more than ever. Against this backdrop, Project Hamilton wisely focused on achieving a *time-to-finality* (also known as *latency*) of fewer than five seconds. TTF "measure(s) the amount of time one has to wait for a reasonable guarantee that crypto transactions executed on blockchain will not be reversed or changed."[29] In other words, it's the time it takes to fully confirm a transaction on blockchain. Confirmation of a transaction on the Bitcoin blockchain, for example, can take about an hour. Other blockchains have faster TTF, but none are instantaneous. Your burrito won't be as hot if you have to wait for confirmation on blockchain before leaving Chipotle and scarfing it down.

Project Hamilton accomplished this mission. A stunning 99 percent of transactions were completed in under two seconds, and the majority took under 0.7 seconds. The project's transactional *latency*—the amount of time between initiating a transaction or payment and receiving confirmation via blockchain—was impressive.

Throughput

Lean manufacturing experts obsess over increasing the number of raw materials they can convert into finished goods. Fifty years ago, the original architects of the internet used packet switching* to maximize the rate at which the nascent network delivered messages.

A close corollary of TTF is *throughput*. In a nutshell, it represents the number of items passing through a system. Project Hamilton aimed for throughput greater than 100,000 transactions per second. After making some technical changes that lay well beyond the scope of this book, it was able to achieve a throughput of an incredible 1.7 million. This is much faster than even the Visa network, which Visa claims is able to process 65,000 transactions per second.[30]

Fault Tolerance

It's no coincidence that Netflix subscribers rarely experience technical problems. Behind the scenes, its engineers have intelligently designed its systems to maximize resilience. When snafus occur and a component fails, more often than not, its systems continue operating normally. This concept is referred to as *fault tolerance*.

To maximize its fault tolerance, Netflix tries to randomly break its own systems. (Yes, you read that right.) The company has even released Chaos Monkey, its sophisticated testing tool, on GitHub for others to freely use and fork.[31]

---

* Put simply, *packet switching* represents a method of breaking data into smaller, digestible chunks and transmitting them over a network. In the end, the message appears in its original form.

For Project Hamilton, the team enabled replication of transaction data over geo-distributed data centers for durability and crash tolerance. All replicas were operated by a centrally administered actor.

Project Hamilton Goes Poof!

Phase 1 of Project Hamilton demonstrated a feasible technical approach, and the researchers promised a Phase 2 that would explore sophisticated approaches to privacy. But in late 2022, the project came under scrutiny from anti-central bank digital currency politicians.

Why? Privacy concerns. These legislators were concerned that a government-issued digital currency would give the government a tool for financial surveillance and control. But, if you recall from Chapter 2, cash is no longer king. Case in point: cash can't even be used at some of our favorite shopping outposts, like Amazon and other online-only retailers.

Furthermore, as discussed earlier in this chapter, debit cards, credit cards, bank account transfers, and apps lack privacy by design. We accept some lack of privacy for the convenience of making electronic payments. Be that as it may, the Boston Fed ended Project Hamilton.

While Project Hamilton is no more, the success of Phase 1 demonstrates that payments with digital currency will work at the required scale when there are sufficient speed, throughput, and fault tolerance requirements.

It's unclear whether the US government will embrace a CBDC. On September 8, 2023, Federal Reserve Vice Chair for Supervision Michael S. Barr stated that the Federal Reserve hadn't yet decided on issuing a CBDC. He indicated that the Federal Reserve would only proceed with the issuance of a CBDC with clear support from the executive branch and authorizing legislation from Congress.[32] However, Vice Chair Barr noted that innovation is important, especially with respect to payments. He said the Federal Reserve

> ... continues to speak to a broad range of stakeholders and conduct basic research in emerging technologies that might support a CBDC payments backbone, or for other purposes in the existing payments system. For example, the Fed's CBDC research program is currently focused on system architecture, notably how ledgers that record ownership of and transactions in digital assets are maintained, secured, and verified, as well as tokenization models—that is, the design of the digital analog to the paper bank note that permits a transfer of value between two parties without direct facilitation by the issuing central bank.

It's unclear whether the Federal Reserve will develop a CBDC and, if it does, whether that CBDC can or should ensure the privacy of those transacting with it.*

CERTAIN PRIVACY CONCERNS OF DIGITAL CURRENCIES

In the coming decades, we'll look back at how digital currencies profoundly affected the way we pay for goods and services. (And by *affected*, I mean *improved*, especially with respect to privacy and security.)

---

* For an interesting perspective on CDBC privacy, check out "The Values of Money: Will Tyranny or Freedom Be in Your Digital Wallet?" by Jim Harper and the Honorable J. Christopher Giancarlo. American Enterprise Institute. February 28, 2023. https://tinyurl.com/2fu3pegr.

Before putting a bow on this chapter, a final note is in order: don't conflate improving user privacy and security with providing *complete* anonymity from governments and authorities. Digital currencies can offer a secure means for the necessary parties to track financial transactions.

Let me be clear: this is a feature, not a bug.

Consider Bitcoin for a moment. It enables parties to conduct transactions using only digital identities. It does *not* require the intervention of a bank, credit card company, or another trusted third party. That's not to say, though, that Bitcoin guarantees anonymity.

Thanks to blockchain analytics, we can connect digital wallet addresses to the people and organizations that use blockchain technology to transact in digital currencies. From one excellent research paper:

> Blockchain analysis is the process of inspecting, identifying, clustering, modeling, and visually representing data on a cryptographic distributed ledger known as a blockchain. The goal of blockchain analysis is to discover useful information about different actors transacting in cryptocurrency. Analysis of public blockchains such as Bitcoin and Ethereum is typically conducted by private companies like Chainalysis.[33]

Yes, terrorists, criminals, tax evaders, authoritarian governments, and other bad actors have used digital currencies for illicit purposes. Along these lines, Andy Greenberg's 2022 book *Tracers in the Dark: The Global Hunt for the Crime Lords of Cryptocurrency* is worth a read. Authorities, however, aren't powerless. Government agencies have used blockchain analysis to track transactions and ultimately bring cybercriminals to justice.

As Robert McMillan writes for the *Wall Street Journal* in April 2023:

> Law-enforcement agencies, working with cryptocurrency exchanges and blockchain-analytics companies, have compiled data gleaned from earlier investigations, including the Silk Road case, to map the flow of cryptocurrency transactions across criminal networks worldwide. In the past two years, the U.S. has seized more than $10 billion worth of digital currency through successful prosecutions, according to the Internal Revenue Service —in essence, by following the money. Instead of subpoenas to banks or other financial institutions, investigators can look to the blockchain for an instant snapshot of the money trail.[34]

Privacy proponents fear that national security and law enforcement will be in conflict with individual privacy. Specifically, they're afraid that applying AML and know your customer (KYC) requirements to all digital currency transactions will force users to reveal sensitive personally identifiable information (PII) such as full names, addresses, birthdates, Social Security numbers, and business partners, all of which is susceptible to hackers. We know this to be true from the examples in the early part of this chapter.

Daniel Gorfine, former chief innovation officer of the US Commodity Futures Trading Commission, and Michael Mosier, former acting director of FinCEN and former deputy chief of the US Department of Justice's Money Laundering and Asset Recovery Section, suggest that we can have our cake and eat it too. In their July 2022 op-ed, they posited that privacy-enhancing techniques (PETs) can create an ideal arrangement.[35] PETs focus on the ability to confirm certain critical information about an individual engaging in a transaction without revealing PII about that individual. PETs include zero knowledge proof, homomorphic encryption, and multiparty computation.

The next sidebar illustrates the use of one particular PET: zero knowledge proof.

Privacy Coin: Zcash

Remember from Chapter 4 that Bitcoin originally offered the promise of anonymous transactions online. As it turned out, though, identifying information for wallet addresses was only pseudonymous. In other words, interested parties could ascertain the identity of its users.

Created and launched by Electric Coin Co. in 2016, Zcash is a privacy-focused blockchain project. It provides enhanced online privacy using something called *zero knowledge proof*. Zero knowledge proof is a novel method by which one party, the prover, can prove to another, the verifier, that a statement is true, without revealing information beyond the validity of the statement itself. It's akin to having someone prove they're over twenty-one without exposing their driver's license information or revealing their birthdate.

It's a revolutionary concept that allows for fully encrypted transactions, ensuring that personal financial information remains confidential. Protecting transactional data can help avoid potential leaks or breaches that can compromise company operations or expose sensitive information.

Chapter Summary

- Current electronic payment methods require users to share a host of personal information that hackers covet.
- Digital currency payments offer greater security and privacy protections or controls than their antecedents.
- The US is just one of many countries currently testing the viability of central bank digital currencies. The results are promising. It's unclear whether the US government will embrace a CBDC. And, if it does, whether it should ensure the complete privacy of those transacting with it like cash.

Chapter 7

Pillar #3: Choice

"Show people a positive path that enables them to make progress on their own terms. Give them options and alternatives that empower them."

— MARK GOULSTON

On September 9, 2014, Tim Cook took the stage at the Steve Jobs Theater in Cupertino, California. Nearly three years after the death of the company's iconic cofounder, Cook was now ensconced in the role of Apple CEO.

Cook wasn't on stage to announce a sexy new piece of hardware à la the iPod, iPhone, or iPad. (The Apple Watch wouldn't arrive for another six months.) Rather, the company had spent years developing a new way for people to pay for things.

As Part I demonstrated, effecting payments is neither simple nor easy. To its credit, Apple's top brass understood this notion exceedingly well. Reporting from the Motley Fool revealed key details of the service's launch:

> Apple's involvement in the world of payments formally began in 2014, but the company had been preparing for its big debut for years beforehand. After acquiring a number of startups and filing patents related to payments, Apple took a leap in 2013 by joining forces with three of the biggest names in the industry: Visa, MasterCard, and American Express.[1]

Apple Pay, Apple's foray into electronic payments, was decidedly on-brand. Over its storied history, the company has demonstrated that it's nothing if not deliberate. Smartphones, MP3 players, smart speakers, and tablets all prove the same point: Apple is *never* the first to market, and Apple Pay proved to be no exception to this rule.

After saying a few words, Cook ceded the floor to Eddie Cue, Apple's senior vice president of internet software and services. Cue told the audience:

> Security and privacy [are] at the core of Apple Pay. When you're using Apple Pay in a store, restaurant or other merchant, cashiers will no longer see your name, credit card number or security code, helping to reduce the potential for fraud. Apple doesn't collect your purchase history, so we don't know what you bought, where you bought it or how much you paid for it. And if your iPhone is lost or stolen, you can use Find My iPhone to quickly suspend payments from that device.

Apple Pay was hardly Apple's first foray into payments.

Let's rewind the clock to June 11, 2012. On that day, Tim Cook announced the launch of Apple Passbook at its Worldwide Developers Conference (WWDC). Passbook launched on September

19, 2012, with iOS 6. After a few years, the company rebranded it as Apple Wallet, releasing iOS 9 on September 16, 2015.

The new moniker was no accident. Even back then, consumers were increasingly using digital wallets. Early entrants like Venmo (circa August 2009) often took the form of mobile apps. These wallets store money, payment credentials, loyalty cards, and other personal and membership information.

They've fundamentally simplified the payment process while concurrently making it more secure and private. They do this by either not storing transaction information on their servers (as Apple does) or using encryption techniques to protect the information (as Venmo does).

But let's get back to Apple Pay. Would it take off? Or had Apple missed the payments boat? Was it too late to overcome the advantage of Android Pay, Google Pay, and others?

Apple Pay skeptics abounded. Writing for CNBC, Cadie Thompson astutely observed:

> One of Apple Pay's limitations is that it is only available to iPhone 6 users for in-store purchases, whereas Google Wallet and other mobile payment apps are available on both iOS and Android.[2]

Those concerns were misplaced; Apple Pay proved to be enormously popular. Figure 7.1 shows its astonishing growth over the five-year period starting in 2016.

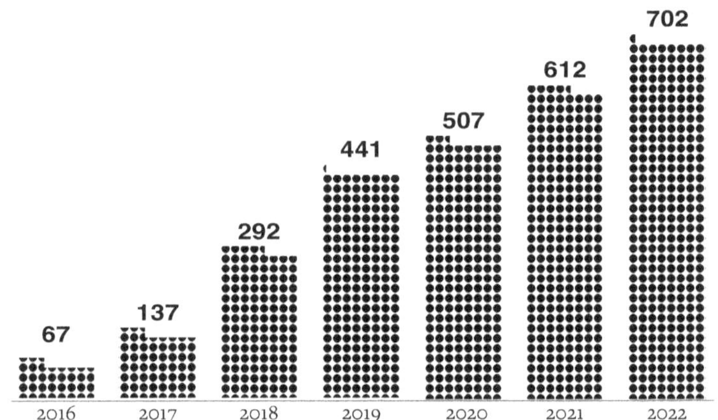

Figure 7.1: Number of Apple Pay Users Worldwide (In Millions)
Source: Statista

And then, of course, COVID happened. In its early days, people were understandably squeamish about touching physical objects, including notoriously filthy bills and coins. Nearly a year into the pandemic, cash transactions worldwide had declined 42 percent from their 2019 level.[3] In 2022, Pew Research found that two in five people don't break out their bills for an entire week.[4] Figure 7.2 shows this increase.

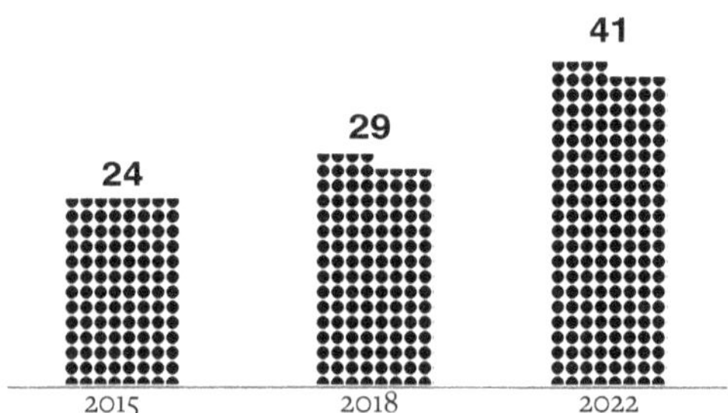

Figure 7.2: Percentage of Americans Who Don't Use Cash for Weekly Purchases
Source: Pew Research

Cash might make a phoenixlike comeback, but I'm not betting on it. Figure 7.3 shows the steady rise in two types of noncash payments: ACH debit and credit transfers.[5]

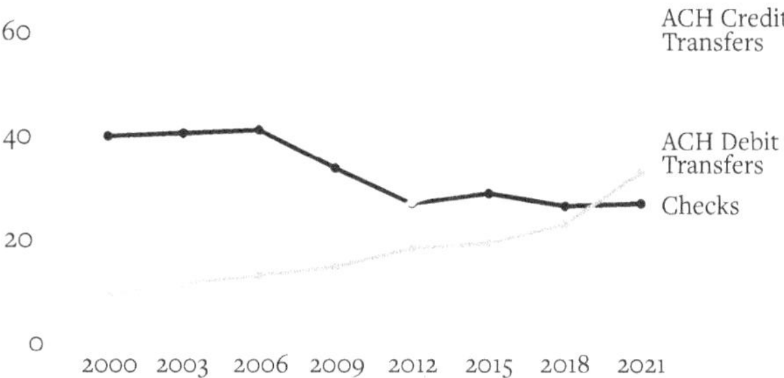

Figure 7.3: 2022 Accessible Version of Trends in Noncash Payments
Source: Federal Reserve

If the use of cash continues to wane, however, it wouldn't be the first time that a popular payment method supplanted an established one.

A Brief Evolution of Payments Over the Past 150 Years

The current popularity of digital wallets underscores two fundamental points. First, consumers have *always* valued the choice in how they pay for goods and services. (Much to the chagrin of the impatient, some people *still* take out their checkbooks at the grocery store.) Second, and in a similar vein, payments have continuously evolved—a point that history bears out.

Let's rewind to post-Civil War America. General stores at the time commonly and willingly let their best customers leave the premises with their arms full of merchandise. In return, store proprietors required nothing more than their customers' IOUs.

If you're struggling with wrapping your head around this concept, search for *Little House on the Prairie* on YouTube. The popular show premiered in 1974 and ran for a decade. *Little House* reflected life in the American Midwest around 1870. Charles and Caroline Ingalls often left the general store—named *Oleson's Mercantile*—with essential supplies and the promise to pay later.

In the pre–credit-card days, merchants handled a good chunk of customer payments on faith. Bob Hunt is the associate director of the Consumer Finance Institute at the Philadelphia Federal Reserve. Payment geeks like me enjoy his working paper, "A Century of Consumer Credit Reporting in America." In it, he writes that nearly a century ago, people financed a remarkable one-third of their retail purchases.[6] Before the Great Depression, sales of goods on credit constituted about one-half of the total store purchases. As discussed in Chapter 2, credit cards wouldn't exist for another two decades.

And general stores weren't alone. Prior to the advent of credit cards, hotel chains, oil companies, and other enterprises would frequently extend informal credit to regular customers.

Fast-forward to 1950 and the arrival of the first credit card. Handshake deals still existed, but plastic started making inroads. Pretty soon, credit cards were on the way to becoming the most popular choice for consumer payments from many socioeconomic backgrounds.[7] Again, the data is compelling.

As of 2022 Americans collectively held almost 530 million credit cards.[8] That year, of US adults, 82 percent had at least one credit card.[9] In 2023, the statistics are even more staggering. The number of credit card holders globally has steadily grown to reach 1.25 billion in 2023, representing an annual growth rate of 2.79% from 1.1 billion in 2018. The United States accounts for 166 million credit card holders. Half hold at least two.[10]

Put differently, credit cards have come a long way from their primitive Diners Club days of the 1950s. For decades, they've been a ubiquitous part of many Americans' lives. (Yet, as discussed in Chapter 5, the unbanked and underbanked are often forced to do without them.)

That's not to say that their run will continue indefinitely. As we've seen with the rise of digital wallets, consumers are already opting for new payment tech.

Digital currency payments may be relatively new, but we've been using high-tech methods and devices to pay for things we need and want for years. As for why, consumers like choice. Within reason, everyone should be able to choose how they pay for things. As we'll see next, payments with digital currencies merely represent an extension of this principle—and, I'd argue, a natural and inevitable one.

New Ways for Consumers to Pay

Although they offer enormous promise, payments with digital currencies are only scratching the surface of the mainstream. In March 2022, ecommerce vendor BigCommerce conducted an extensive consumer survey. Only 5 percent of consumers are *currently* using digital currencies when shopping online. Interestingly, consumers "would be more willing to pay with cryptocurrency if brands offered it as an option."[11] Put differently, payments with digital currencies may well appeal to more than just a niche audience.

When it comes to payments, consumers have *always* appreciated choice. Progressive organizations smell opportunity and dollar signs. To this end, some have already started accepting digital currency payments.

BURRITOS AND BITCOIN

Let's play a quick game of word association.

Chipotle.

What's the first thing that comes to your mind? Go.

The odds are high that you didn't say *digital currency*. Interestingly, few retail outlets have embraced payments with digital currencies as much as the fast-casual Mexican chain.

In April 2021, Chipotle commemorated National Burrito Day by giving away $200,000 in free burritos and digital currencies.[12] Here's a little more on the promotion:

> Chipotle said this week that it's giving away $200,000 in cryptocurrency through a new online game called *Buy the Dip*—a nod to the crypto market's ongoing downturn. Playable through July 31, the game gives players a chance to win crypto like Bitcoin, Ethereum, Solana, Avalanche, and Dogecoin along with other prizes that are actual dips such as guacamole and queso.[13]

Prizes broke down as follows:

- Ten thousand fans won one free burrito.
- Fifty fans won $500 in Bitcoin.
- Three won $25,000 in Bitcoin.

But Chipotle didn't stop with *Buy the Dip*. Chief Marketing Officer Chris Brandt knew he could use other ways to fuse his company's offerings with gamification and digital currencies. Why not keep experimenting and pushing the envelope?

A year later, Chipotle unveiled Burrito Bucks, its own "in-experience currency" in collaboration with the online game platform Roblox. Burrito Bucks accompanied the launch of Chipotle's *Burrito Builder* game, also on Roblox. Players could exchange the currency for 100,000 free entrée codes at participating Chipotle restaurants. Thousands of customers rolled their own virtual burritos and won free in-store wraps.[14] Figure 7.4 shows an image from the game.

Figure 7.4: *Burrito Builder* Game
Source: Roblox, Chipotle

A mere thirty minutes after launching *Burrito Builder*, people gobbled up all 100,000 codes. At its peak, 27,000 people concurrently created virtual burritos.[15]

Then, in June 2022, Chipotle began another innovative campaign. All US-based stores started accepting popular digital currencies as payments.

In the interest of full disclosure, Chipotle works with my current employer, Flexa. As a result of the partnership, its customers can use about 100 different digital currencies, such as Bitcoin, Ethereum, Avalanche, Solana, and Dogecoin to pay for their food at Chipotle.[16]

Flexa's president and cofounder Trevor Filter described the partnership as follows:

> Chipotle benefits from reaching customers where they are through the new payment methods they want to use. And, at the same time, Chipotle's customers benefit from greater payment flexibility, faster checkout, and a more seamless experience at the fast-casual restaurants they already love.[17]

In September 2022, Flexa and Chipotle teamed up again to commemorate the successful Ethereum Merge.* Beginning on September 15, 2022, Chipotle and Flexa offered an exclusive "proof of steak" savings on Chipotle orders featuring the Garlic Guajillo Steak. When customers used Flexa to spend ETH at any Chipotle restaurant in the US, they saved 99.95 percent off their purchase. This was the amount equal to the projected overall energy savings from Ethereum's transition to a proof-of-stake consensus mechanism. The promotion was supposed to run

* The Ethereum Merge occurred when the original Ethereum Mainnet proof-of-work blockchain merged with a separate proof-of-stake blockchain called the Beacon Chain. The Merge reduced Ethereum's energy consumption by ~99.9 percent.

through October 4, 2022; however, due to overwhelmingly positive response, the promotion ended early on September 29, 2022.

Digital currency skeptics might be asking, so what? Other than PR, did the promotions yield any *real* benefits?

RESULTS

Brandt certainly thinks so. By his estimation, the last digital currency giveaway brought in nearly 4 million unique visitors who played 26 million times. Nearly 7 million people played the Chipotle *Burrito Builder* Roblox game tens of millions of times. As Brandt explains, "We certainly want to cement ourselves with fandom from the newest generation of both Gen Z and beyond and people who are tech-savvy and digitally savvy."[18]

Moreover, the foray into games has served as an effective recruiting tool for Chipotle's massive loyalty program. As of July 2022, Brandt claims it sports a stunning 29 million members.

There's no cash in the Web3 Metaverse. Chipotle's Roblox and Flexa partnerships banked on the growing popularity of Web3, consumer desire for gamification, and customer interest in a new payment choice. It worked.

Balancing Increased Choice With Security and Other Obstacles

As we've seen in this chapter, new payment methods are promising and exciting. Their manifold benefits include:

- Faster transaction processing time.
- Lower transaction fees.
- Increased liquidity.
- Strengthened security.

It's essential to balance these benefits with drawbacks. Payments with digital currencies pose challenges. All are solvable, but a few words about them are in order.

CHALLENGES

Consumers may like choice, but they like security and privacy too. Consider the words of Beth Costa and Neeko Gardner, two payments experts at the consulting firm Oliver Wyman. As they astutely observe in "How Gen Z Shops and Pays":

> The adoption of cryptocurrency as a payment method is minimal today, but this is an area that should be monitored for potential growth and more widespread use going forward. Younger generations have a much lower tendency to use credit cards and [are] willing to use alternative payments, opening themselves to new ways of engaging with merchants that offer payment options integrated into their daily lives.[19]

Generally speaking, digital natives are more open to new technologies than their older counterparts. Next, as you'd expect, payments made with digital currencies can introduce a fair degree of uncertainty. Consider merchants who accept digital currency payments. Blockchains are immutable and irreversible. Thus, when a consumer returns a good that she paid for with digital currencies and the merchant can't reverse the blockchain transaction, the merchant must find a solution for a refund. (Don't worry, we discuss this concept in Chapter 10.)

Finally, the volatility of digital currencies is a significant issue that buyers and sellers face. It requires the merchant to immediately convert the digital currencies to fiat currencies. That's unless, of course, a stablecoin pegged 1:1 with a fiat currency is used for the transaction or the merchant holds the digital currency on its balance sheet.

TOO MUCH CHOICE?

Pandemic-driven supply-chain anomalies aside, American consumers have long lived in a sea of choice. Globalization, advances in shipping, and the rise of ecommerce have collectively increased the availability of goods and services. But is there too much choice?

That very question intrigued psychologists Sheena Iyengar from Columbia and Mark Lepper from Stanford University. In 2000, the two published a now-famous study about—of all things—jams.[20]

Iyengar and Lepper set up a table at a local food market. One day, they displayed twenty-four different kinds of jams for purchase. What's more, they offered samples to interested folks. The next morning, they did the same thing, but they stripped their table down to only six flavors.

The results were remarkable. While the twenty-four-jam table generated more interest than its six-jar counterpart, passers-by were about ten times more likely to buy a jar from the table with far fewer options. It turns out that less is often more when it comes to choice.

As NYU professor and serial entrepreneur Scott Galloway has put it, "Consumers don't want more choice, they want to be more confident in the choices presented."[21] (For more on this subject, check out Barry Schwartz's excellent 2004 book, *The Paradox of Choice: Why More Is Less.*)

Looking Forward: Payments, Choice, and the Metaverse

It's easy to find digital currency fans and metaverse fans. But how far off is widescale adoption? It's hard to tell. However, consumers show a growing willingness to explore shopping in the metaverse and paying for virtual items with digital currencies. Over one-quarter of consumers have a strong understanding of the metaverse. Nearly half are willing to shop in the metaverse and purchase both virtual and physical goods. For example, the luxury retailer Balenciaga launched a collection of virtual outfits and accessories in the game *Fortnite*. Shoppers could purchase limited-edition Balenciaga x *Fortnite* clothes in-store and online.[22]

Gen Z and millennials are the most willing to shop on the metaverse, while Gen X and baby boomers are the least willing. Willingness to shop in the metaverse scales as income grows.

The main reason consumers don't use digital currencies to pay for goods and services is because they don't understand it. However, another top reason is because retailers simply don't offer it as a payment method. Gen Z and millennials understand digital currencies better than their older counterparts. Once brands begin to offer digital currencies as a payment option, it's likely that the number of consumers who use it will grow.

RALPH LAUREN EMBRACES NFTS

Count Walmart CTO Suresh Kumar among the bulls. In October 2022, he predicted that digital currencies would "become a major payments disruptor."[23] David Lauren, chief branding and innovation officer at the retailer Ralph Lauren, is another bull.

Lauren is clearly putting its money where its mouth is with the launch of digital currency payments and a Web3 experience at its Miami outpost.

In April 2023, Ralph Lauren announced that its new Miami Design District store—the first Ralph Lauren store to do so—would accept purchases with digital currencies including Bitcoin (BTC), Ethereum (ETH), and Polygon (MATIC). Ralph Lauren also announced a partnership with the Web3 community Poolsuite to gift NFTs.* The first element of the partnership was a co-designed Ralph Lauren x Poolsuite NFT that was gifted to approximately 3,000 Poolsuite community members. The NFT unlocked access to a party at a private waterfront Miami estate that was part of a three-day launch event.[24]

To my knowledge, the gambit represents Ralph Lauren's first proper blockchain-based venture. Expect the company and others to find innovative ways to reach customers wherever they are. (Yes, that includes online gaming and the metaverse.)

PREPARING FOR THE METAVERSE

Will Chipotle and Ralph Lauren ultimately reap the benefits of the high-tech experiments in Web3, the metaverse, digital currencies, and NFTs? There's good reason to believe so.

Consider some of the results from PwC's February 2023 *Global Consumer Insights Pulse Survey*:

- Twenty-six percent of survey participants said they've participated in metaverse-related activities in the past six months for entertainment, virtual experiences, or purchasing products.

---

* Chapter 4 introduced this subject.

- Nine percent of respondents said they've purchased an NFT.

One thing is certain: no one takes cash in the metaverse. There are opportunities to increase customer bases in the metaverse, and cracking the payments nut there will reap rewards.

Chapter Summary

- Payments have never been static. Smartphones and other technological advances have spurred the development of digital wallets and payment apps.
- Antiquated payment rails cost us time and money.
- Digital currencies represent the next step in the evolution of payments. Major retailers like Chipotle and Ralph Lauren are already on board.
- The metaverse represents the next payment frontier where opportunity abounds and cash isn't accepted.

Chapter 8

Pillar #4: Efficiency

"There's a way to do it better. Find it."

—THOMAS EDISON

Over Labor Day weekend in 1995, Pierre Omidyar listed a broken laser pointer on his nascent site AuctionWeb. His expectations for the auction were low, and he set the initial bid at $1.

For a week, there was nary a sole bid on the busted device, but then something happened: a few people did. Ultimately, Mark Fraser purchased it for $14.83.[1] Astonished, Omidyar contacted the buyer to ensure he knew the pointer was unusable. Fraser did know, but that fact didn't deter him from the purchase. Fraser couldn't afford the cost of a new laser pointer at the time (~$100) and, as an electrical engineer, he believed he could fix Omidyar's broken one.

Ever the contrarian, Omidyar became increasingly convinced that people would purchase a wide variety of items from complete strangers online.

Plenty of folks at the time thought that Omidyar was foolish, but user data indicated otherwise: AuctionWeb was taking off. A few years later, Omidyar rebranded his company as eBay, and its initial public offering (IPO) took place on September 24, 1998. The rest, as they say, is history. (For more on the company's astonishing tale, check out Cohen's excellent 2002 book *The Perfect Store*.)

Payments, Pauses, and Painful Processes

It's nearly impossible to tell the entire eBay story, however, without mentioning the considerable friction and inefficiencies that its users routinely experienced in its early days. A simple example will illustrate the, er, painful process from a quarter-century ago.

EARLY EBAY: NOT-SO-SIMPLE TRANSACTIONS

Let's stay in the mid-1990s for a moment. *Friends* fans like me will appreciate this example. Let's say that Rachel lists her lava lamp on eBay. Joey bids on the auction and ultimately submits the winning bid. Here's the list of steps for how the transaction might be consummated:

1. Rachel sends an electronic message to Joey with the final amount owed, including shipping. She also includes her physical address.
2. Joey writes a check to Rachel and mails it.
3. Rachel and Joey wait for the check to arrive at Rachel's home.
4. Six days later, Rachel receives the check.
5. The following day, Rachel deposits the check at her local bank.
6. Rachel waits for the funds to clear.

7. Upon receiving confirmation that the bank has released the funds, Rachel sends Joey the lava lamp from her local post office.
8. A week later, Joey finally receives the package.

This hypothetical transaction takes two weeks to complete, assuming nothing gets lost in the mail.

Chuckle all you want at this primitive way of conducting commerce, but it was the norm for eBay buyers and sellers for years. As late as July 2002:

> 60 percent of the $13 billion in merchandise sold annually on eBay is paid by check or money order, and sellers usually wait for the checks to clear before shipping their items out—a process that can take two weeks.[2]

This process didn't exactly represent the acme of efficiency.

CONFINITY AND THE PROMISE OF ELECTRONIC PAYMENTS

Max Levchin, Peter Thiel, and Luke Nosek, tech-savvy entrepreneurs, were among the first to grasp the enormous future impact of the web. The three found the possibilities endless and, in December 1998, founded Confinity, a start-up intent on enabling "low-cost, almost effortless digital payments for consumers and businesses."[3]

Interestingly, the troika was far more knowledgeable about—and comfortable with—programming than payments. Consider the words of Russel Simmons, an early über-talented recruit:

> It's kind of funny in retrospect because we didn't know anything about payments ... we had never written code that had

> interacted with a database. ... We didn't know well enough to know that we should be more intimidated by the problem.[4]

About a year later, Confinity launched its flagship product, PayPal.

The product quickly took off, and Confinity's momentum accelerated. The company merged with Elon Musk's payments start-up X.com in March 2000. Confinity soon rebranded as PayPal. On February 14, 2002, it began trading on Nasdaq as a public company. (For more on this story, see Jimmy Soni's 2022 book *The Founders: The Story of PayPal and the Entrepreneurs Who Shaped Silicon Valley*.)

EBAY REVISITED

By this point, eBay's efforts to build its own proprietary infrastructure for digital payments weren't bearing sufficient fruit, and the pressure to innovate was intensifying. Despite scant profits, Amazon was establishing itself as an ecommerce juggernaut. eBay execs at the time feared the potential impact of the 1999 launch of Amazon Auctions. (As it turned out, those concerns were misplaced. Amazon ultimately shuttered its eBay clone in 2001.[5])

Against this backdrop, Omidyar and his team began exploring potential acquisitions. It didn't take long to bag the elephant. On July 8, 2002, eBay announced that it was acquiring PayPal. As Brian Bergstein of the Associated Press wrote at the time:

> Online auction giant eBay Inc. is buying Internet payment provider PayPal Inc. in a $1.3 billion stock deal that would unite arguably the Web's most successful business with one of the few companies that has been giving it any trouble.[6]

The story of eBay and PayPal illustrates our fourth and final pillar: making and receiving payments should be as efficient and frictionless as possible. Put differently, no one should have to wait longer than necessary to transact business. Payments with digital currencies can significantly increase efficiency and should, therefore, lower costs for all parties involved.

Payment Problems: Why the Status Quo Is Suboptimal

Whether you're whipping out your credit or debit card at the counter or online, rest assured: payments within our current financial system generally work. That's not to say, though, that the status quo is above reproach. On the contrary, the current regime is less than ideal. Let's explore why.

NOT SO SWIFT: ADVENTURES IN CROSS-BORDER PAYMENTS

Some payments are simpler to remit than others. For example, a Wisconsin hotel contracts a local, third-party caterer to supply the hors d'oeuvres for a small tech conference. The accounts-payable clerk can match the invoice number to the purchase order. After that, she can manually cut a check and pay the caterer's invoice without much of a hassle. Payments that cross international borders, however, can be much trickier.

Of course, none of this is news. The gold standard for cross-border payments, SWIFT, traces back to 1973. That year:

> 239 banks from 15 countries got together to solve a common problem: how to communicate about cross-border payments. The banks formed a cooperative utility, the Society for Worldwide Interbank Financial Telecommunication, headquartered in Belgium.[7]

Think of SWIFT as primarily a communications service that acts as an interbank messenger. For example, a user need only pay a small fee and provide their bank with the following information:

- The SWIFT code to the seller's bank.
- The seller's account number.

This information allows the user's bank to securely send information regarding the payment over the SWIFT network to the counterparty's bank. When the counterparty's bank receives a notification of the impending payment, it credits the money to the counterparty's account. The payment generally takes a few days to reach its destination. However, longer waiting periods may occur for a variety of reasons, including when "intermediary banks" are involved, as described below.

In effect, SWIFT allows banks to send important messages that facilitate payments.

Like most things from half a century ago, the process tends to work, but no one today would consider it efficient.

Inefficiencies with SWIFT stem from the system's reliance on human beings. In the SWIFT universe, say that Banks A and B have been conducting business for two decades. Regardless of the type of transaction taking place, Bank A still needs to confirm that it's making a payment to Bank B. Once Bank A has received the requisite confirmation from Bank B, it kicks off the manual process of transferring the funds.

What happens if one of the banks isn't a SWIFT member? In those cases, SWIFT routes funds through intermediary banks. "Adding more parties to a transaction speeds things up," said no one ever.

The SWIFT network works, but it's slow during a time when people are fundamentally impatient. (A 2015 survey found that 96 percent of Americans knowingly consume hot food or beverages that burn their mouths.[8]) That impatience extends to getting their money, receiving confirmations, and the like. And here's where the status quo suffers.

Transactions on SWIFT's rails sometimes incur unexpected fees. What's more, delays are common. As of 2019, SWIFT's self-reported error rate clocked in at 6 percent.[9] Dismiss that number if you like, but consider the following question: would you be happy if that same percentage of your emails required additional human intervention to receive them?

SWIFT's Blockchain Experiment

On August 31, 2023, SWIFT announced the results of a series of experiments demonstrating that its infrastructure can seamlessly facilitate the transfer of tokenized value across multiple public and private blockchains.[10] SWIFT focused on delivering instant and frictionless cross-border transactions.

Why did SWIFT focus on tokenized assets? Well, even though tokenization of assets hasn't yet been widely adopted, 97 percent of institutional investors believe it will revolutionize asset management and be a positive force in the financial industry.[11]

SWIFT leveraged its network and the Chainlink Cross-Chain Interoperability Protocol (CCIP) to create an experimental solution. SWIFT collaborated with more than a dozen financial institutions and financial market infrastructures—Australia and New Zealand Banking Group Limited (ANZ),

BNP Paribas, BNY Mellon, Citi, Clearstream, Euroclear, Lloyds Banking Group, SIX Digital Exchange (SDX), and The Depository Trust & Clearing Corporation (DTCC)—to discuss technical and nontechnical considerations that would need to be addressed to make a proposed solution commercially feasible.

The goal was to test whether SWIFT could enable financial institutions to use their existing back-end systems to interact with tokenized assets and transact across both public and private blockchain platforms.

The results of the experiment were compelling. SWIFT demonstrated a secure, scalable way for financial institutions to connect to multiple types of blockchains.*

PAYING TO PAY

Payments made with credit, debit, and prepaid cards are nearly ubiquitous. To be sure, these methods are convenient, but they come with additional costs. Yes, I'm talking about a slew of different fees. Writing for *Forbes*, Kimberlee Leonard and Cassie Bottorff provide an excellent summary of them,[12] which Table 8.1 details.

* This wasn't SWIFT's first foray into blockchain and digital currencies. In 2021, SWIFT published a white paper that assessed the potential impact of CBDCs and how SWIFT could support the financial community as new currencies are developed. In 2022, SWIFT tested how its network could be leveraged to support the seamless integration of both CBDCs and tokenized assets into the existing financial system. In early 2023, SWIFT tested a new CBDC interlinking solution with eighteen central and commercial banks in a sandbox environment.

Type of Fee	Description
Interchange fees	Payment made directly to the card issuer for the swiped transaction. Fees may vary based on the type of card used, the transaction amount, and the industry the business operates in. For example, credit card companies often charge higher interchange fees for online purchases because fraud represents a more significant risk with these types of transactions.
Payment processor fees	The payment processors referenced in Chapter 2 often charge fees for facilitating transactions. Merchant-services charges include monthly fees, per-transaction fees, equipment lease fees, and statement fees. Processors make their money this way; they collect none of the aforementioned interchange fees.
Assessment fees	Paid directly to the credit card network so the merchant can accept certain credit cards. These fees are based on total monthly sales; they're not transaction based. When combined with the interchange fee, merchants refer to the total as a swipe fee.

Table 8.1: Types of Credit Card Processing Fees

Let's put these fees in context. Total credit card processing fees for businesses typically break down as follows:

- **Interchange fees:** Range from 1.15 percent + $0.05 to 3.15 percent + $0.10 for Visa and Mastercard, from 0.05% to 1.75%, with a per-transaction fee of $0.04 to $0.25 for debit cards, and 1.4% to 3.15% for American Express.[13]
- **Assessment fees:** An additional 0.14 percent to 0.17 percent in assessment fees.[14]

To state the obvious, fees vary. In 2022, the average US per-transaction fee for credit and debit cards was 1.7 and 0.9 percent, respectively. Businesses pass these fees along to consumers. As of April 2021, only five states have prohibited the practice: Colorado, Connecticut, Kansas, Maine, and Massachusetts.[15]

Whether consumers notice these fees or not, make no mistake: they're considerable. Unfortunately, they're also on the rise.

Nilson is one of the most trusted sources of global news and statistics about the payment industry. From its 2021 report:

> Credit, debit, and prepaid cards issued in the United States, general purpose, and private label type combined, generated $9.443 trillion in purchase volume in 2021, up 23.2% from 2020. US merchants who accepted those cards as payment for goods and services paid $137.83 billion in processing fees, an increase of 24.3% from the prior year.
>
> ... merchants paid $32.60 billion to accept debit and prepaid cards. This was an increase of 21.7% over 2020. Debit and prepaid cards accounted for 23.7% of the total processing fees merchants paid in 2021. This was a decline from 24.1% the prior year.[16]

In 2022, Visa and Mastercard announced plans to increase their fees for fund features like fraud prevention, innovation, and rewards programs. Due to pressure from US legislators, they delayed the start of the increases to the fall of 2023 and the spring of 2024. The increases apply to a variety of transaction types, including many types of online purchases as well as in-store purchases in grocery stores and other retail settings.

The consequences of increased fees are real. In 2019, the FDIC found that high processing fees cause many households to remain unbanked or underbanked.[17] (As we saw in Chapter 5, lack of banking is a real problem with profound consequences.)

Bottom line: massive opportunity exists to reduce the exorbitant cost of making payments. Doing so could save billions of dollars—in many cases, by the people least equipped to endure these fees.

LACK OF SPEED

Checks take time to clear. Ditto for money orders and bank transfers. In some cases, it may take several days before recipients can access available funds. Payments made with credit and debit cards allow people to walk away with their purchases faster. But speed comes at a cost.

FRAGMENTATION AND LACK OF INTEROPERABILITY

Let's say you want to send money to a friend or family member in another country. In 2023, you'd think that the process would be relatively simple. Sadly, however, it's not.

Unfortunately, payment systems in one country don't seamlessly integrate with their counterparts in another country. (That's why SWIFT is experimenting with blockchain.) These systems often rely upon their own standards for storing data and payment-related messages. As a result, effecting an international payment often requires *interlinking arrangements*. The Bank for International Settlements defines these as:

> ... a set of contractual agreements, technical links and standards, and operational components between payment systems of different jurisdictions, allowing their respective participating PSPs [payment service providers] to transact with one another as if they were in the same system.

> PSPs historically and to this day generally make cross-border payments to PSPs in different payment systems by using the services of local agents, or correspondent banks, in those payment systems. In a correspondent banking arrangement, the correspondent bank holds deposits owned by a bank in a foreign payment system (the respondent bank) and provides those respondent banks with payment and other services.[18]

Kristalina Georgieva is the managing director of the International Monetary Fund, or IMF. In her words, the international payment system consists of "the financial roads, railways, bridges, and tunnels that allow currencies to be exchanged and capital to flow between countries."[19] It's an apt metaphor. The current system includes a labyrinth of:

- Links between correspondent banks.
- Messaging systems, such as SWIFT.
- Money transfer businesses.
- Credit card networks.
- Foreign exchange markets.
- Arrangements between central banks.

If this system sounds complex, trust your judgment. And, as you'd expect, consumers pay for this complexity. Much like with payments made with credit and, to a lesser extent, debit cards, the cost of remitting a cross-border payment is high. Consider the following data from the World Bank (WB):

> Sending remittances comes at a cost that can be rather steep, depending on the type of service used to move money across borders. The global average in Q3 2021 was 6.3 percent, more

> than double the Sustainable Development Goal (SDG) target of 3 percent of the amount sent. According to the WB, banks were the most expensive service provider, with a global average cost of 10.40 percent. That is $20.80 in fees remitters needed to pay for sending $200 to their family back home.[20]

Remitting money doesn't need to be this way—nor should it. The same holds for making purchases.

NOT JUST CONSUMERS: REIMAGINING COMMERCIAL PAYMENTS

In this chapter, I've primarily focused on the benefits consumers can expect by embracing payments with digital currencies. (Management types refer to these transactions as business-to-consumer, or *B2C*.) I'd be remiss, however, to ignore the other side of the ledger.

Digital currency payments also present massive opportunities for the business-to-business (B2B) realm. Allied Market Research estimated that global B2B payments reached $125 trillion in 2021. By 2031, that number may exceed a whopping $313 trillion.[21]

With so much on the line, you may reasonably assume that organizations have, by and large, eliminated inefficient, manual payment processes. Nothing could be further from the truth. In 2020, McKinsey estimated that:

- An astonishing 30 percent of all B2B payments rely upon bank transfers.
- Nearly half of all global business transactions still require physical paper.[22]

One word: wow.

Explaining the Status Quo

With so much inefficiency and legacy tech, it's natural to wonder: why is this still the case?

Allow me to posit two reasons.

US-SPECIFIC ISSUES

In the United States, start-ups in the payment space face an uphill battle. The credit card networks have an entrenched multi-party ecosystem that makes it hard for newcomers to succeed. In other countries, however, start-ups have gained significant traction. In China, for instance, digital wallets and super apps came to market soon after the launch of iPhones and Android smartphones. Within a few years, China had simplified and revolutionized digital payments:

> As low-cost, open-sourced Android smartphone adoption began to ramp up in China in 2012, ecommerce gained traction on Alibaba mobile apps. In what was largely a cash economy, Alibaba recognized the need for a trusted solution that could scale at a low cost for mass consumer adoption to easily pay for ecommerce payments online.[23]

Let's not overstate things, however. It's unfair to place all the blame for inefficient payments at the feet of the current financial and banking systems. And this brings us to the second reason: corporate inertia.

CORPORATE INERTIA

Although many organizations have automated their internal and back-office processes over the past twenty years, plenty remain stuck in the twentieth century. Companies tend to cling to legacy systems to pay vendors, partners, and suppliers.

No company is an island, though. Even those that use state-of-the-art payment technologies and systems exist in a suboptimal reality.

It's time to talk about better ways to make payments.

Solutions: New Tech and Payment Methods

The opportunity to significantly improve the status quo is vast. Relying upon mid-twentieth-century technology in 2023 is unwise. Today there are ways to more quickly, affordably, and securely route funds as needed. Consumers and businesses stand to benefit a great deal from payments with digital currencies. As for how, approaches vary, but here are some of the most exciting projects currently underway.

FEDNOW

In July 2023, the Federal Reserve launched the FedNow service. Think of it as a new instant payment service that enables consumers and businesses to settle payments nearly instantaneously.

This is accomplished through deposit accounts with banks that maintain a master account at a Federal Reserve Bank. Although the FedNow service currently supports only domestic payments between US depository institutions, the service could eventually facilitate cross-border payments if access expands to non-US financial institutions in the future.

NOT YOUR FATHER'S BLOCKCHAIN: TASSATPAY GETS CREATIVE

Startups like New York-based Tassat Group Inc. use blockchain technology to make commercial banking transactions more efficient, faster, and secure. Tassat created a private, permissioned blockchain-based payments platform for the banking industry and its B2B customers called TassatPay.

TassatPay's solution enables payments across member banks with The Digital Interbank Network, a private, permissioned blockchain-based payments network.* The Digital Interbank Network is composed of FDIC-insured banks transacting real-time payments and performing other banking services between their corporate clients.† TassatPay uses Google's global network, leveraging a single Virtual Private Cloud, and Google Cloud Load Balancing to optimize web access for customers across multiple back-end instances and multiple regions.

Not surprisingly, TassatPay's results have been impressive. The company has processed more than $1 trillion in transactions as of February 2023.[24]

It eliminates the pain points associated with legacy bank payment channels, or rails, and enables banks to provide their B2B customers with secure, real-time transactions 24/7/365 using blockchain technology.

---

* A permissioned blockchain is a distributed ledger that's not publicly accessible. Only users with permissions can access it.

† Tassat Group has developed more than twenty use cases, including logistics, mortgage warehousing, commercial construction, private equity capital calls, and broader working capital applications for banks' corporate clients.

LIGHTSPARK

If we can *instantly* stream movies and send emails, why can't we move money as fast?

Answering that question primarily involves solving a scalability problem. For example, the Bitcoin network, which was created as a peer-to-peer electronic payment system, struggles to settle large amounts of transaction data in a short period.

Well aware of this problem, Joseph Poon and Thaddeus Dryja formally proposed a solution in their 2016 paper.[25] Their Lightning Network is an open-source, layer 2 payment protocol* built on top of Bitcoin.[26] However, users have reported liquidity problems and security weaknesses, creating a perception that the Lightning Network can be complicated and problematic.[27]

Enter Lightspark Group Inc. (Lightspark), a company that has built enterprise-grade infrastructure to make it easier for businesses to connect to the Lightning Network for low-cost and instant payments.†

David Marcus is Lightspark's founder, and his bona fides around digital payments are impressive. Marcus is the former president of PayPal and the cocreator of Facebook's ill-fated Libra/Diem stablecoin project. According to Marcus:

> We believe the Internet badly needs an open payment protocol—one that works 24/7, settles in near real-time, is dirt cheap, interoperable, and open to all to build on ... Money should move online like emails or text messages, and

* For more on the layer 2 payment protocol, see https://tinyurl.com/repay-layer2.

† For more about Lightspark, check out www.lightspark.com/.

> the Lightning Network has the best chance of becoming the standard protocol that enables that and much more for everyone around the world.[28]

Will Marcus and Lightspark ultimately succeed? The answer hinges upon consumer adoption. As Tomio Geron astutely points out on *Protocol*:

> Ultimately, consumers will pay with whatever's most convenient. Merchants may show interest in cheaper payments, since they typically bear the cost, but wide acceptance and consumer interest [are] crucial. It's going to be a technological race between Lightning, which can leverage the best-known cryptocurrency, and all the other crypto payments contenders.[29]

Lightspark's commitment to simple, reliable, intuitive, and secure payments holds great promise.

PIX

Roberto de Oliveira Campos Neto is a Brazilian economist and the current president of the country's Central Bank. As he told *Reuters* in November 2020, "Huge changes are underway in payments. Society demands something that is fast, cheap, safe, transparent, and open."[30]

Neto isn't just speaking off the cuff. In that month, Brazil's central bank launched a revolutionary payment system. Dubbed Pix, the instant payments platform promised to:

> ... speed up and simplify transactions, as well as foster financial sector competition and lure in new players such as big techs Facebook and Google.

> The state-owned instant payments system allows consumers and companies to make money transfers 24 hours a day, seven days a week, without requiring debit or credit cards. It is also free of charge for individuals.[31]

Pix enables its users—people, companies, and governmental entities—to send or receive payment transfers in a few seconds at any time, including on nonbusiness days. Pix is über-modern, efficient, and simple. The payer asks for the Pix alias or scans a QR code (static or dynamic) to start a Pix funds transfer on any day and at any time.

Pix tends to have a lower acceptance cost for merchants and businesses in general because its transactional framework has fewer intermediaries than traditional methods of payment.

Within a year, half of the country's population used Pix—roughly 110 million Brazilians.[32] The number has since grown to 120 million.[33] Figure 8.1 displays its popularity compared to other payment apps.

Brazil Central Bank data indicates that 35 percent of transactions in Brazil are currently made through the Pix system. By the end of August 2023, individual Pix users numbered 153 million.[34]

Although Pix wasn't built using blockchain technology, its success had garnered the attention of many digital currency projects and companies, including the US digital currency exchange Coinbase. In March 2023, Coinbase introduced a Pix integration.[35] It partnered with Brazilian payment solutions provider EBANX and Pix to facilitate easy purchasing of digital currencies using the Brazilian real, as well as enabling deposits and withdrawals from Coinbase accounts in real.

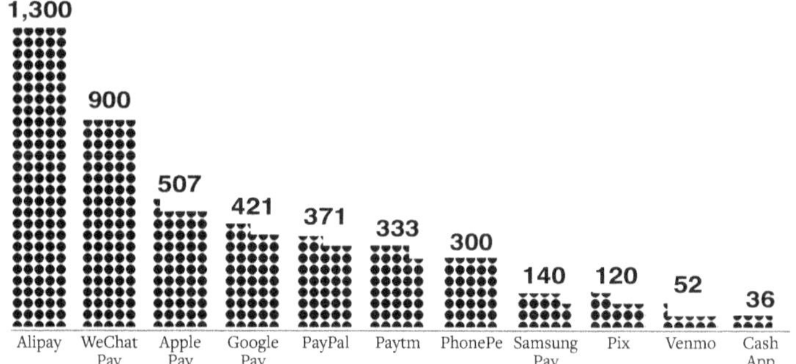

Figure 8.1: Number of Users for Popular Payment Methods (In Millions)
Source: Non-Pix Data: Fintech News as of September 22, 2022, Pix Data: UTORG

RIPPLE FACILITATES CROSS-BORDER PAYMENTS

As we discussed in Chapter 2, cross-border payments are typically made through a global network of correspondent banks and involve multiple intermediaries that are fragmented across different time zones and operating hours. They exhibit high costs, low speed, operational complexities, limited access, and low transparency. This causes inefficiencies and settlement risk that are ultimately detrimental to end users.

In 2011 developers Jed McCaleb, Arthur Britto, and David Schwartz started building a better way to facilitate cross-border payments "in seconds not days."[36] Their efforts ultimately spawned RippleNet (used for making international money transfer between banks) and XRP Ledger (a decentralized public blockchain that allows for the fast, low-cost, real-time transfer of XRP, fiat currencies, and other digital assets).

Advantages Galore

In many ways, RippleNet represents an improvement over SWIFT. A fascinating case study in *Harvard Business Review* reveals its significant advantages:

> Access to RippleNet allowed payment providers and banks to send and receive real-time payments with payment tracking, pre-transaction validation of fees, as well as the ability to include invoices and other rich payment data. Network participants could confirm payment details before initiating transactions to ensure settlement certainty, minimizing settlement risk and failures.[37]

RippleNet offers financial institutions an even more important benefit.

Liquidity

Ruth Porat currently serves as the CFO of Alphabet, Google's parent company. In 2020, Forbes listed her as the sixteenth most powerful woman in the world. Before joining Google as chief financial officer, she rose to the same position at Morgan Stanley.

Porat knows a few things about money matters. As she once famously said, "Liquidity is oxygen for a financial system."[38]

She's right.

Liquidity represents an organization's "ability to convert assets to cash or acquire cash—through a loan or money in the bank—to pay its short-term obligations or liabilities."[39] Without sufficient liquidity, companies can quickly fail, as can entire economies.

RippleNet has allowed payment providers and financial institutions to reduce liquidity costs using a digital currency created by Ripple called XRP. Real-time access to capital eliminates the need for prefunded accounts. As Ripple CEO Brad Garlinghouse told HBR:

> There's trillions of dollars parked around the world, prefunded in between banks, between corporates, corporates at banks, and if we can make that—those trillions of dollars—more efficient, we make the entire global financial ecosystem more efficient.[40]

Viamericas, MercuryFX, Cuallix Bank, and Catalyst Corporate Federal Credit Union are just a handful of financial institutions that have chosen to utilize RippleNet. SWIFT may not become obsolete anytime soon, but alternative payment networks that use digital currency and blockchain-based protocols to effect cross-border payments are here to stay. Expect more to come.

It's not just private companies that are facilitating better cross-border transactions, as you'll read about next.

PROJECT MBRIDGE

Project mBridge is a joint project between the Swiss-based Bank of International Settlements Innovation Hub Hong Kong Centre and four participating central banks in Asia and the Middle East: the Hong Kong Monetary Authority, the Bank of Thailand, the Central Bank of the United Arab Emirates, and the Digital Currency Institute of the People's Bank of China.

Project mBridge is a blockchain technology pilot built to support real-time, peer-to-peer, cross-border payments and foreign exchange transactions using CBDCs. It ensures "compliance with jurisdiction-specific policy and legal requirements, regulations, and governance needs."[41]

mBridge is important because central banker technologists themselves designed the blockchain protocol, calling it the mBridge Ledger, or mBL. Each of the four central banks involved in the pilot operates a node on the mBL platform to validate all transactions.

The six-week pilot represented the largest cross-border Central Bank Digital Currency pilot to date. More than $12 million was issued on the platform, facilitating over 160 transactions worth more than $22 million.[42]

Part II explained how digital currencies can improve payments for all concerned. Part III assumes that you want to get started or, at a minimum, that you're at least curious about exploring the next steps. To ensure a successful launch, it's time to focus on how to accept these payments. I'll concentrate on operational considerations.

Chapter Summary

- We used to think nothing of saying, "The check is in the mail." Advancements in payments, however, mean that that mindset no longer suffices.
- Payments made using the current banking and financial systems often suffer from high costs, fragmentation, and interoperability issues.
- As a result, the status quo is ripe for disruption with blockchain and digital currencies.

- These new technologies knock down existing barriers and give way to faster, more affordable, efficient, and secure ways to make payments.

Part III:

From Theory to Practice: Making Digital Payments Happen

Chapter 9

Regulation and Strategies

"It is Congress' job to regulate this industry."

—KIRSTEN GILLIBRAND

In July 2023, I flew down to Washington, DC. No, I wasn't going to see a Nationals game or check out the Smithsonian. Rather, I attended an intimate dinner with Representative Maxine Waters.

The California Democrat has served as the US representative of her state's forty-third congressional district for more than three decades. Many view her as one of the most powerful women in American politics today. Waters has gained a reputation as a fearless and outspoken advocate for women, children, people of color, and the poor.

Most relevant for our purposes, Congresswoman Waters made history as the first woman and first black chair of the House Financial Services Committee in 2019. (After the Republicans won the House in 2020, Waters became the HFSC's ranking member. In other words, she's the most senior member of the committee from the minority party.)

The HFSC is an important animal in American politics, banking, and finance. From its website, the committee maintains:

> jurisdiction over issues pertaining to the economy, the banking system, housing, insurance, and securities and exchanges. Additionally, the Committee also has jurisdiction over monetary policy, international finance, international monetary organizations, and efforts to combat terrorist financing.[1]

I found Waters to be engaging, approachable, and convivial. (Fun fact: Tupac Shakur was her favorite rapper.) She eagerly listened to different points of view, and I enjoyed our discussion.

This meeting wasn't my first visit to Capitol Hill, though. Over the past several years, I've regularly met with congressional leaders tasked with critical financial matters. Our discussions have focused on digital currencies, blockchain, and payments.

From these conversations, it's clear that wholesale regulation of digital assets is forthcoming because of the proliferation of the asset class. Senior company leaders who ignore this reality do so at their own peril.

This brief chapter accomplishes two goals. First, it describes the evolving regulatory landscape, with a particular emphasis on the US. Second, it offers several high-level strategies on how organizations can embrace payments with digital currencies while successfully navigating this complex regulatory framework.

Disclaimer

The regulatory environment around digital currencies and payments is dynamic. No book—never mind a chapter—could possibly cover all the digital currency legislation in the US and abroad. Rather, this chapter gives an overview of key trends and possible legislation.

Regulation of Digital Currency Payments

Imagine if organizations could remit money to other groups, individuals, and governments willy-nilly—even ones that trafficked drugs, women, and children and financed terrorism. Silk Road would look tame by comparison.

It wouldn't be good for anyone.

US PAYMENT LAWS

Most industrialized nations have passed laws regulating the safe and secure transmission of funds. The US is no exception to this rule. The federal government and forty-nine of the fifty states have enacted regulations to oversee entities that engage in the business of money transmission. (The lone holdout: Montana. Don't ask me why.)*

Legal definitions of a *money transmitter* vary, but a simple definition follows: an individual or business that engages in the

* Notwithstanding the foregoing, on March 10, 2023, Montana issued a letter stating, "Money transmitters must email detailed business plans exactly describing their proposed activities in Montana and flow of funds structure at every point in the transaction (including who holds the funds and how) to the Division of Banking and Financial Institutions..."

transmission of money or value. What does that mean exactly? Well, it's the receipt (or the facilitation of the receipt) of money or value from person or company A and its transmittal to person or company B.* Businesses such as Square, Venmo, Flexa, and PayPal are all examples of licensed money transmitters.

Federal Landscape

At the federal level, the Bank Secrecy Act (BSA) of 1970 requires money services businesses, which include money transmitters, to register with the Financial Crimes Enforcement Network (FinCEN). The BSA is an anti-money laundering (AML) statute that aims to prevent money laundering, terrorist financing, and other illicit crimes. A key component of compliance with the BSA is to know your customer (KYC). KYC is a mandatory process of identifying and verifying the identity of the client or customer when she opens an account at a financial institution or money services business operating in the US.

Money transmitters are also required to comply with economic and trade sanctions imposed through executive orders and congressional legislation. This is to protect US economic interests, foreign policy, and national security. Sanctions programs are listed in the *Federal Register* and administered through the Office of Foreign Assets Control (OFAC), an agency of the United States Department of the Treasury. Money transmitters

---

* Money transmission can involve the receipt of money or value from a person or company in one location and its transmission to that same person or company at a different location. An example might be the intermediary who receives money from my TD bank account and transmits it to my Coinbase account so that I can purchase digital currencies from the platform.

must ensure that the funds that pass through their hands aren't remitted or received in violation of OFAC regulations. They do this by performing due diligence in checking the Specially Designated Nationals and Blocked Persons List (SDN) and blocking or rejecting prohibited transactions as required by law.

With respect to digital currencies, since 2011, FinCEN has issued interpretative guidance related to the asset class.* It was the first US financial regulator to address and assign obligations to digital currency money services businesses to guard against financial crime. In 2018, OFAC added digital currency addresses to the SDN to alert the public of specific digital currency identifiers associated with a sanctioned person.

State Landscape

As previously mentioned, forty-nine of the fifty states have enacted regulations to oversee entities that engage in the business of money transmission. State requirements typically include maintenance of a license in a state, including being subject to examinations by state regulators (which are typically at the business's expense), and submission of financial audit reports, background and character examination for key executives, surety bonds, and minimum net worth or collateral requirements.

---

* FinCEN uses the term "convertible virtual currency" instead of digital currency used here. According to FinCEN, "convertible virtual currency" is a medium of exchange that operates like a currency in some environments because it either has an equivalent value in real currency, or it acts as a substitute for real currency. But it doesn't have all the attributes of real currency; for example, convertible virtual currency generally lacks legal tender status in a jurisdiction. FIN-2013-G001, "Application of FinCEN's Regulations to Persons Administering, Exchanging, or Using Virtual Currencies." FinCEN (March 18, 2013).

These requirements are designed to protect consumers. Compliance can be onerous and time-consuming for the business.

State laws aren't uniform, unfortunately. In an attempt at uniformity, the Conference of State Bank Supervisors (CSBS), with input from a working group consisting of industry participants and regulators, developed the Model Money Transmission Modernization Act (Model Act). Since the beginning of 2023, twelve states—Arkansas, Georgia, Hawaii, Indiana, Iowa, Minnesota, Nevada, New Hampshire, North Dakota, South Dakota, Tennessee, and Texas—have passed legislation based, to varying degrees, on the Model Act.

For digital currency payments, a number of states have addressed regulation of these activities under their money transmission laws. But state money transmission regulators haven't found a consistent way to apply their laws to digital currency activities. New York and Louisiana require a specific license to conduct money transmission involving digital currencies. California recently enacted similar legislation. New Jersey is considering passing similar legislation as well. Over a dozen states have addressed the regulation of digital currencies through amendments to their money transmission statutes. Other states have issued interpretive guidance. A significant number of states still haven't established a formal, public position on the regulation of digital currency activity.

OTHER RELEVANT US LAWS

The US has a complex web of overlapping laws and regulations that apply to digital currencies. Federal and state-level agencies treat the asset differently based on its function in a transaction. For example, the Internal Revenue Service (IRS) treats digital currencies as property for tax purposes. The Commodity Futures Trading Commission (CFTC) classifies them as commodities, whereas the Securities and Exchange Commission (SEC) classifies them as securities.

Many banking and financial statutes have been on the books for years—and sometimes decades. The rise of digital currencies and high-profile fraud cases (read: FTX), however, have underscored the need to pass legislation targeted directly at digital currencies.

TRENDS DRIVING DIGITAL CURRENCY REGULATION

Although it's not a comprehensive list, what follows are the major trends forcing politicians and regulators to the table.

Industry Growth

Saying that the global market for digital currencies has exploded is a gross understatement. On May 1, 2013, Bitcoin's market capitalization clocked in at a mere $1.3 billion.[2] Fast-forward seven years. Bitcoin remains the 800-pound gorilla, but the collective value of more than 5,100 different digital currencies exceeded $231 billion.[3] (And that's *after* the most recent crypto winter resulted in massive haircuts to the values of just about all remaining digital currencies.)

Blockchain technology, on which digital currencies are built, continues to flourish. Financial services, payments, asset management, art, music, and gaming are just a few of the areas that stand to benefit from blockchain technology and digital currencies.

The Incumbents Want In

Large banks, brokers, and other financial institutions have had digital currencies and blockchain on their radar for years. For example, in 2018 Goldman Sachs launched GS Accelerate, a small in-house incubator that encouraged employees to examine innovative uses of digital currencies.[4] Other firms and consultancies did the same in fear of the costs of inaction. That is, if they're not devoting significant resources and bandwidth to these areas, they run the risk of appearing hidebound.

Over the past five years, digital currencies have matured, and some countries have passed related legislation. (More on this subject in a bit.) Against this backdrop, organizations are increasingly loath to sit on the sidelines. Case in point: in October 2021, PayPal launched a new service enabling users to buy, hold, and sell digital currencies.[5] Then on August 7, 2023, PayPal announced the launch of a US dollar-denominated stablecoin, PayPal USD (PYUSD).[6]

The publicly traded digital currency exchange Coinbase, in partnership with The Block, recently published "The State of Crypto: Corporate Adoption." The report aptly notes,

> [Fortune 500] companies, among the world's largest and best known, are innovating and investing in [crypto, blockchain, and web3] technologies because they know that our century-old global financial system needs updating, that blockchain can be a foundational solution, and that not keeping pace will mean losing ground in this global economy to competitors around the world, among other reasons.[7]

The report found the following among survey respondents,

- More than half—52 percent—of the Fortune 100 companies have pursued digital currency, blockchain, or Web3 initiatives since the start of 2020.
- About 60 percent of Fortune 100 companies initiatives reported since the start of 2022 have been in the prelaunch stage or have already launched.
- 83 percent of surveyed Fortune 500 company executives who are familiar with digital currency or blockchain say their companies either have current initiatives or are planning them.
- The companies with the highest number of initiatives consist of four of the largest tech companies, four of the largest banks, a retail giant, and a beverage colossus.
- The top 10 Fortune 100 companies in volume of Web3 initiatives are: (1) IBM (18 initiatives); (2) Alphabet (11 initiatives); (3) Microsoft (11 initiatives); (4) Goldman Sachs (10 initiatives); (5) JPMorgan Chase (9 initiatives); (6) Amazon (6 initiatives); (7) Citigroup (6 initiatives); (8) Coca-Cola (5 initiatives); (9) Nike (5 initiatives); and (10) Bank of America (5 initiatives).

These companies are taking a cue from MicroStrategy, a publicly traded maker of business intelligence software, which we explore in Chapter 10. In August 2020, MicroStrategy began acquiring large quantities of Bitcoin. For diversification, companies have started holding digital currencies on their balance sheets. Some of the largest and most prestigious college endowments, like Harvard, Yale, and Brown, have bought Bitcoin. Banks are using digital currencies to collateralize loans, with more examining the prospect every day.[8]

There's also an increasing popularity of digital currency investment products—in particular, Bitcoin investment products. Investment firms and asset managers are increasingly allowing clients to invest in digital currency through a variety of financial instruments, such as digital currency-focused funds. On September 14, 2023, Deutsche Bank announced it would offer (via a partner) custody services for institutional clients' digital currencies and tokenized assets.[9]

But clearer regulatory rules in the US are essential for mass adoption of digital currencies.

The Decline of TerraUSD

Recall from Chapter 4 that price-based stablecoins maintain price stability relative to a national currency. At least that's the theory.

Terra was once a prosperous blockchain network with its own non-fungible tokens (NFTs), decentralized finance (DeFi) platforms, and Web3 applications. But then two of its ecosystem tokens, LUNA and TerraUSD (UST), caused its collapse. UST is a stablecoin with a price pegged to the US dollar, meaning that one UST should be valued at $1. This value was achieved with UST's sister token, LUNA, using a smart contract–based algorithm that would burn (eliminate) LUNA tokens to mint new UST tokens.

UST's $1 peg was maintained until May 7, 2022, when the price of the stablecoin began to dip below $1. It was officially de-pegged from the dollar two days later, on May 9, after falling to $0.35.[10] In another few days, on May 12, the plunge continued. The price of LUNA dropped to $0.000017.[11]

Officials, including US Treasury Secretary Janet Yellen, have urged Congress to regulate stablecoins, noting the "risks to financial stability."[12]

Ouch. This collapse pales in comparison, however, to an event that occurred a few months later.

The FTX Implosion

In November 2022, the digital currency exchange FTX collapsed.* News outlets were all atwitter, even comparing the scale of the implosion to Enron. (Check out Michael Lewis's book *Going Infinite: The Rise and Fall of a New Tycoon.*)

FTX was a digital currency exchange that focused primarily on buying and selling digital currency derivatives. It was available in certain US states, the Bahamas, Japan, Europe, Switzerland, and Hong Kong. Aside from its own token that the exchange developed and sold, FTT, FTX allowed for the trading of Bitcoin, Ethereum, XRP, Tether, and more than 300 trading pairs.†

The fall of FTX and its founder, Sam Bankman-Fried, can be traced back to the liquidity crisis of the FTT token and Bankman-Fried's trading firm, Alameda Research (Alameda). Bankman-Fried had moved up to $10 billion in FTX customer assets to minimize losses by Alameda. Alameda's assets were primarily locked up in the FTT token.

Then Changpeng "CZ" Zhao, the founder and CEO of the

* See https://tinyurl.com/ftx-collapse for a complete timeline of events.

† Trading pairs are assets that can be traded for each other on an exchange. Examples are Bitcoin/Litecoin (BTC/LTC) and Ether/Bitcoin Cash (ETH/BCH).

digital currency exchange Binance, announced that Binance would liquidate its stake in FTT, which led to an increase in customer withdrawals of FTT. Ultimately, it led to the bankruptcy of FTX.

Bankman-Fried was later arrested. On November 2, 2023, he was convicted of fraud and other related crimes. He faces up to 110 years in prison.

The collapse of FTX left an estimated 1 million creditors facing losses of billions of dollars. It demonstrates the importance of transparency, investor protection, and regulatory clarity for the asset class. It proves that terrible things can happen when a single, centralized group that lacks adequate risk management procedures like FTX controls exchanges, wallets, and market-making services. It also accentuates the need to separate customer assets from corporate ones because Bankman-Fried used FTX customer assets to minimize the losses of his other company, Alameda.[13]

FORTHCOMING US REGULATION

Against this backdrop, US Congress members have introduced several bills over the past year, including:[14]

- **Responsible Financial Innovation Act:** Introduced by US Senators Cynthia Lummis (R-WY) and Kirsten Gillibrand (D-NY), this bill creates a comprehensive regulatory framework for digital assets to protect consumers, prevent fraud and abuse, and create transparency and accountability in the digital asset marketplace.
- **Digital Asset Anti-Money Laundering Act of 2022:** Introduced by US Senators Elizabeth Warren (D-MA) and

Roger Marshall (R-KS), this bill is intended to mitigate the risks that digital assets pose to US national security by bringing the digital asset ecosystem into greater compliance with anti-money laundering and countering the financing of terrorism rules that govern the rest of the financial system.

- **Digital Commodities Consumer Protection Act:** Introduced by US Senators Debbie Stabenow (D-MI) and John Boozman (R-AR), this bill gives the CFTC the authority to regulate the trading of digital commodities—mandating consistent, rigorous rules for all market participants.
- **Digital Asset Market Structure Proposal:** Introduced by Patrick McHenry (R-NC) and Glenn "GT" Thompson (R-PA), this bill provides a statutory framework for digital asset regulation intended to provide clarity, fill regulatory gaps, and foster innovation, while providing adequate consumer protections.
- **Clarity for Payment Stablecoins Act:** Introduced by Patrick McHenry (R-NC), this bill establishes a framework for regulation, supervision, and enforcement of stablecoin issuers.

To be sure, fundamental differences exist among these proposed pieces of legislation—and more will invariably come down the pike. Some seek to formally grant the Securities and Exchange Commission jurisdiction and enhanced authority over digital currencies and related activities. Others would authorize the Commodity Futures Trading Commission to do the same. As of this writing, none has made it to President Biden's desk for signature.

The lack of regulatory clarity is a barrier to digital currency payments in the US. In a recent survey conducted by Coinbase, 46 percent of the executives say that it's one of the biggest barriers to adoption.[15] A whopping 91 percent of the respondents believe that the uncertain regulatory environment makes it tricky to navigate the digital currency space. And 92 percent of the executives think that policymakers should enact new legal frameworks that promote new technologies instead of stifling innovation with old rules meant for older technologies.

US politicians debate, analyze, and listen to arguments about digital currencies, but the world keeps turning. As of this writing, America has fallen behind other countries fostering innovation in this key area.

Many international governments have recognized the need to create clear rules, guardrails, penalties, and oversight for digital currencies. The next section looks at some examples of international regulatory regimes.

EUROPEAN UNION

For years, the European Union has been working on the passage of the Markets in Crypto-Assets Regulation (MiCA), a compressive digital asset framework applicable to all twenty-seven EU member states.[16] Specifically, MiCA uses the term *crypto assets*, which it defines as "digital representation[s] of value or rights which may be transferred and stored electronically, using distributed ledger technology or similar technology." It differentiates between the following three types of crypto assets, which are referred to as tokens:

- **Utility token:** A type of crypto asset that's only intended to provide access to a good or a service that its issuer supplies.
- **Asset-referenced token (ART):** A type of crypto asset that's not an electronic money token and that aims to maintain a stable value by referencing another value or right or a combination thereof, including one or more official currencies.
- **E-money token (EMT):** A type of crypto asset that purports to maintain a stable value by referencing the value of one official currency.

MiCA endeavors to enhance transparency, protect investors, and establish a comprehensive framework for the asset class. MiCA will replace the disjointed laws and regulatory environment currently in place in the EU. Once MiCA comes into effect, EU member states must ensure that their local laws align with MiCA's framework. The European Securities and Markets Authority is the primary regulator with the authority to license and regulate entities.

The good news: companies that must comply with MiCA will be able to operate across all EU member states after obtaining a license from a regulator in one EU country.

In addition to MiCA, the EU adopted the Transfer of Funds Regulation (TFR). It's anti-money laundering legislation requiring those who engage in digital currency transfers to collect identifying information on the parties involved in the transaction, similar to customer identification procedures used in traditional fund transfers. This is to prevent, detect, and investigate money laundering and terrorist financing.

Certain MiCA regulations will take effect on June 30, 2024, and the remainder will take effect on December 30, 2024.

UNITED KINGDOM

In April 2022, HM Treasury, the UK government's economic and finance ministry, announced its intention to promote digital asset technology innovation. Then Chancellor of the Exchequer, Rishi Sunak, who became prime minister in October 2022, said:

> It's my ambition to make the UK a global hub for cryptoasset technology, and the measures we've outlined today will help to ensure firms can invest, innovate and scale up in this country.[17]

In February 2023, the UK government announced its plans for robust regulation of crypto asset activities. The UK aims to provide clarity and protection for consumers and businesses through licensure, disclosure requirements, data reporting, consumer protection, and operational resilience.

The first set of regulations will adapt existing traditional finance frameworks to digital currency exchanges, custodians, and other financial intermediaries. Entities will need to obtain a license, meet capital/liquidity requirements, and comply with the Financial Conduct Authority's (FCA) custody rules.*

In June 2023, UK parliamentarians voted through the Financial Services and Markets Bill (FSMB) that would "recognize crypto as a regulated activity in the country."[18] In October 2023, HM Treasury announced that legislation for fiat-backed

* The FCA regulates the financial services industry in the UK. Its role includes protecting consumers, keeping the industry stable, and promoting healthy competition between financial service providers.

stablecoins would be introduced in early 2024. Regulation in other areas, such as algorithmic stablecoins, will follow as the government brings activities like lending and trading with digital currency into the fold of conventional financial regulation.

It's too soon to tell if FSMB will strike a balance between managing potential consumer and asset stability risks and creating an environment in which digital asset providers can operate, innovate, and grow.

SINGAPORE

Singapore, led by its central bank, the Monetary Authority of Singapore (MAS), has been at the forefront of regulation of digital assets. Since 2019, it has worked to build a clear regulatory framework, including a robust licensing regime, anti-money laundering regulations, and consumer and investor protection regimes.

Depending on its characteristics, each digital asset may be treated as a regulated product such as a capital markets product (including securities), e-money, a digital payment token (DPT), or an unregulated digital token that's used only for utility purposes.

Singapore passed the Payment Services Act (PSA), a framework for payment systems and payment service providers, in 2019. The PSA requires payment service providers to obtain a payment license from the MAS. Payment service providers are those that engage in the following services: account issuance; e-money issuance; cross-border money transfer; domestic money transfer; merchant acquisition; DPT; and money-changing. For example, entities such as digital currency exchanges and those that issue and facilitate payments in stablecoins must obtain a license for regulated payment services.

MAS also licenses virtual asset service providers (VASPs) that apply and meet certain requirements under the PSA.

In January 2022, MAS issued guidelines prohibiting digital currency businesses from advertising in public areas, including on buses, ATMs, and public websites and in broadcast and print media. These businesses can, however, market services on their own websites, mobile apps, or social media accounts.

In March 2022, MAS set out anti-money laundering and terrorist financing standards, consistent guidance issued by the Financial Action Task Force (FATF).

FATF

FATF is a global anti-money laundering and terrorist financing watchdog. It has thirty-nine members, including the European Union, the United States, the United Kingdom, Saudi Arabia, and Hong Kong.

It promotes global standards to mitigate risks associated with money laundering, terrorist financing, drug trafficking, arms trading, cyber fraud, and other serious crimes.

FATF also assesses whether countries are taking effective action to prevent and combat such crimes.

UNITED ARAB EMIRATES

The Middle East is swiftly developing frameworks for digital asset businesses aiming to become innovation-centric economies that support the growth of the ecosystem, including the metaverse and Web3.[19]

In March 2022, the Virtual Assets Regulatory Authority (VARA) was established in the Emirate of Dubai to regulate all activities related to the digital assets sector. Less than a year later, in February 2023, VARA issued a regulatory framework governing digital assets in the Emirate.

The regulations include supervision and enforcement powers of VARA. The regulations are applicable to virtual asset service providers (VASPs), which are companies involved in everything from issuance, exchange services, and custody to brokerage and advertising services.

VASPs must be licensed by VARA and comply with, among other regulations, the Compulsory Rulebooks: Company, Compliance and Risk Management, Technology and Information, and Market Conduct. They include provisions for custody and segregation of client money, insurance and liquidity reserves, anti-money laundering, know-your-customer, travel rule considerations, market manipulation/abuse prevention, data privacy, and information security.

VASPs also need to comply with activity rulebooks corresponding to their specific services, including advisory, broker-dealer, custody, exchange, lending and borrowing, transfer and settlement, and management investment.

High-Level Strategies

It's a good bet that wholesale federal regulation is coming to the US at some point. That doesn't mean, however, that organizations need to (or should) wait until legislators and regulators finalize new regimes. This chapter concludes with four high-level strategies to accept customer payments with digital currencies.

DO NOTHING—AT LEAST FOR NOW

As we saw in Part II, a few benefits of enabling digital currency payments include the ability to reach new customers, lower transaction fees, and offer consumers greater choice.

However, uncertain regulation in the US, growing—but not wholesale—demand from customers to use digital currencies as a form of payment, and the volatile nature of the asset class provide legitimate reasons to sit back and wait.

On the other hand, inaction has its own set of risks. If your competitors are incorporating digital currency payments, your company could find itself at a competitive disadvantage.

MONITOR AND INVESTIGATE

This cautious approach involves paying attention to everyone:

- Federal, state, and international legislative and regulatory developments.
- Your competitors.
- Your customers.

Although this approach fails to provide the vaunted first-mover advantage, its potential drawbacks aren't as daunting.

LAUNCH A PILOT PROGRAM

In Chapter 7, we saw how Ralph Lauren's Miami location let customers make payments with digital currencies, but it's hardly alone. Every day, more businesses take the plunge. Even more telling, the notoriously slow public sector is starting to dip its toe in the water.

Along these lines, consider this fascinating announcement from October 2022:

> Colorado is accepting crypto as payment for any taxes owed to the state as of Sept. 1. It was the result of a promise made earlier in the year by Colorado Governor Jared Polis, who has proven his commitment to establishing the state as pro-cryptocurrency.[20]

Stay tuned. More public and private entities are recognizing that the benefits of accepting payments with digital currencies are too big to ignore.

GO ALL IN

Finally, an increasing number of organizations have ripped off the bandage and gone all in on digital currencies.

Equinox NYC Clubs

Equinox is a chain of high-end athletic clubs, with a particularly strong footprint in Manhattan. *Fitness Magazine* even voted it the best gym in the nation. As of May 2022, members have been able to use digital currencies to pay for their annual memberships.[21]

Sheetz

With nearly 700 convenience stores spread across Pennsylvania, Maryland, Virginia, West Virginia, Ohio, and North Carolina, Sheetz is no mom-and-pop outfit. Yes, the challenges of accepting payments in multiple states are considerable, but that didn't stop the chain from partnering with Flexa to allow customers to "pay for items inside the store or fill up their cars, trucks, and RVs at the pump using digital currencies like Bitcoin, Ether, Litecoin,

Dogecoin, and more."[22] In so doing, Sheetz beat 7-11, Circle K, Wawa, and others to the punch.

Shopify

In 2020, the ecommerce behemoth started partnering with digital currency payment providers to enable merchants to accept digital currencies as a payment method.[23] In 2022, Shopify supported more than 4 million merchants in 175 countries.[24] In August 2023, Solana Pay, the payment protocol built on the Solana blockchain discussed in Chapter 4, created its own Shopify payments app, allowing users to pay in USD Coin (USDC).[25]

Expect more companies to follow suit, but how exactly does a company legally accept payments with digital currencies?

Let that query serve as the starting point for the next and final chapter.

Chapter Summary

- Digital currency payments are generally subject to federal and state laws used to oversee entities that engage in the business of money transmission.
- More than ever, the US needs clear digital currency regulation; other countries are much further along in developing wholesale regulatory regimes for the asset class.
- Industry growth, desire for incumbent payment and financial institutions to offer digital currency payments, and extreme instances of fraud and mismanagement by certain projects have enhanced the need for US legislation and regulation.

- Companies are taking different approaches to accepting digital currency payments, from inaction to monitoring to jumping into the deep end.

Chapter 10

Execution Considerations

"Vision without high-impact execution is hallucination."

—HENRY FORD

MotorCars of Georgia is a luxury auto shop in Atlanta. Customers visit the shop to buy McLarens, Aston Martins, Bentleys, and other luxury vehicles.

In the fall of 2017, a thirty-five-year-old techie named Peter Saddington walked into a MotorCars location near his home. Ultimately, he purchased a 2015 Lamborghini Huracan for $200,000.

Ho hum, right? It's tempting to dismiss the story as another luxury car purchase. But here's where things get interesting.

Saddington had bought Bitcoin in November 2011 for $2.52 per token. Over time, he purchased more, resulting in nearly $200,000 in gains, which he used to purchase the Huracan.

To complete the transaction, however, Saddington didn't just hand the salesperson a bunch of digital currency—not that that's even possible. Rather, he sold some Bitcoin and then used the fiat proceeds to buy his dream car.

As you'd expect, the story went viral. CNBC's *Make It* interviewed Brandon Saszi, the general manager of MotorCars of Georgia. In his words:

> "I've been in this business long enough that nothing really surprises you. The only thing you're thinking is, 'Gosh I wish I was in on it.'"

But what if Saddington—or anyone else for that matter—*didn't* have to take that intermediate step of liquidating the Bitcoin to purchase the car? That is, what if MotorCars accepted Bitcoin as a form of payment as Chipotle started doing in 2022?

What if all merchants and businesses followed suit and accepted not only Bitcoin but other digital currencies as well? As discussed throughout this book, that trend is catching on.

So, how would an enterprise make this happen? Where would they even begin?

Let those two questions serve as the North Star for this book's final part. This chapter focuses on operational issues.

Enabling Digital Currency Payments

In the Preface, I discussed how the dot-com bust provided cover for many old-school, brick-and-mortar executives intent on ignoring the massive consumer convenience that ecommerce provided. Regardless of *when* an individual brick-and-mortar enterprise finally embraced online shopping, it needed to implement new technologies, change some existing business practices, adopt new ones, and educate customers.

As we'll see in this chapter, accepting payments with digital currencies requires extensive cross-functional organizational involvement. In other words, it takes a village. Let that simple, yet powerful, statement serve as the introduction to this section.

Truth in Advertising: A Disclaimer

This tactical chapter provides a high-level overview of important concepts and technologies around accepting payments with digital currencies. Think broad, not deep. For a slew of reasons, it's not feasible to provide an entirely comprehensive list of steps to begin this journey.

One size has never fit all, and it never will.

Getting Started: An Overview

Benjamin Franklin famously said, "By failing to plan, you are preparing to fail." He wasn't talking about buying burritos or luxury cars with Bitcoin, but his wise words certainly apply here. At a minimum, any organization beginning its journey needs to ask and answer the following questions:

- Which digital currencies will the business accept as payment for transactions?
- What should the business expect in terms of processing time and fees for transactions?
- How will returns and refunds be processed?
- Is it best to jump into the deep end or just dip a toe in the water?
- What, if any, are the legal or regulatory requirements in the jurisdictions in which the business operates?

- What types of partners, technologies, and integrations should a business use to accept payments with digital currencies?
- Will the business hold digital currencies on its balance sheet?

I could go on, but you get my point: quite a few variables are at play.

Before addressing these meaty and essential questions, a few notes are in order. First, answering them will take time.

Second, they involve all parts of the organization. Thinking of integrating digital currency payments as an IT or finance project is a grave mistake. Yes, those folks need to be involved, but it's also critical to involve other company stakeholders such as those in the engineering, marketing, HR, legal, and operations departments.

Finally, the answers to the questions in this chapter may change over time. The business, fiscal, social, technological, political, and regulatory environments are fluid.

Technology

The current payments landscape is complicated. Recall the discussion in Chapter 2 of the various steps and parties that facilitate payment authentication, clearing, and settlement services. These payment systems often provide the plumbing for payment transactions and are relevant to accepting digital currency payments. Here we discuss some of the main factors that curious organizations must consider.

ACCEPTING PAYMENTS

A company can accept payments in digital currencies like Bitcoin and Ether in two main ways:

- **Do it yourself:** The company sets up a digital currency wallet to accept, store, and use digital currency for its business from customers. Customers pay for goods and services by transferring digital currency directly from their wallet to the company's wallet. (Chapter 6 discusses this topic in far more detail.)
- **Use partners:** The company partners with various third-party service providers, which could include wallet providers, exchanges, custodians, payment gateways, and payment processors. These third parties facilitate the company's receipt of digital currencies or fiat for payment of goods or services and help the company—generally using existing hardware and software—with payment acceptance.

SOME CONSIDERATIONS FOR SELF-MANAGED DIGITAL CURRENCY PAYMENTS

For option 1, which we will call *self-managed digital currency payments*, there are a host of considerations for deployment. First, the company will pay fewer transaction fees because of the lack of partners and intermediaries in the flow of funds. The second consideration relates to security. Private keys secure the digital currency assets in the wallet. A "seed phrase" (generally at least twelve words) is created every time a user opens a new digital currency wallet. Think of a seed phrase as a master key for your digital currency. If the company loses access to its digital currency wallet, a seed phrase can help recover the digital currency. However, this means that if somebody steals the company's seed phrase, they gain access to the company's digital currency assets. The company must, therefore, be exceptionally safe with its seed phrases.

Third, the payment will settle in whatever digital currency the customer used for the transaction. If the company wants to convert the digital currency into traditional fiat money, it must use a third party, such as a digital currency exchange, to facilitate this process. (See the next section in this chapter titled "The Technical Scaffolding Behind Accepting Digital Currency Payments.")

Fourth, self-managed digital currency payments enable companies to accept digital currency payments from all over the world. The downside is that the company could unknowingly engage with illicit actors who use digital currency transactions to engage in fraud, launder money, finance terrorism, or evade sanctions. This can create civil and criminal liability for a company. The company should ensure that it does anti-money laundering, know-your-customer, and sanctions screening on customers. Companies like Coinbase offer "self-managed commerce" products that allow a company to control its private keys, but Coinbase ensures that the company doesn't engage in transactions with sanctioned individuals.

Fifth, unless a company uses a new receiving address for each transaction (incurring additional network fees to sweep funds for settlement), anyone—including competitors—will be able to view the volume of their transactions on-chain.

Sixth, blockchain transactions are immutable and irreversible, so a company must consider how, if at all, it will facilitate returns and refunds.

THE TECHNICAL SCAFFOLDING BEHIND ACCEPTING DIGITAL CURRENCY PAYMENTS

Whether a company decides to go it alone or work with others to accept digital currency payments, both will rely on some or all of the following technologies.

Application Programming Interfaces

At its core, an application programming interface, or API, is a software program that allows different applications, systems, and web services to easily talk to each other. With respect to payments, APIs are powerful, flexible, and—not surprisingly today—practically ubiquitous.

Consider Stripe, one of the world's largest payment processors. The company allows organizations "to accept a variety of payment methods through a single API."[1]

When it comes to payments, APIs provide for vast, constantly expanding possibilities. For example, some APIs allow developers to add blockchain payment integration to a company's existing online checkout process.

Code Libraries

As a lot, coders generally hate duplicating their efforts. Why reinvent the wheel? JavaScript, Python, and other general-use programming languages rely heavily on powerful libraries for a slew of purposes, including data visualization, analysis, web development, machine learning, and image recognition.

With respect to digital currencies, code libraries allow developers to focus on payment flow and ecommerce integration. That is, developers need not worry about mundane but critical specific details when building their wares.

Website Payment Buttons and Plug-Ins

Next, let's consider website payment buttons. Paying on a website or mobile app with PayPal has been *de rigueur* for years. The same thing is happening with digital currencies.

When it comes to evaluating technology partners, know this: the technology built by third-party vendors should easily and seamlessly integrate into your organization's existing in-store payment systems, website, and ecommerce software. If that's not the case, ask yourself whether you identified the right vendor or if you need to overhaul your company's technology, which could be antiquated. Antiquated websites, point of sale (POS) terminals, customer relationship management (CRM), and enterprise resource planning (ERP) systems may thwart efforts to embrace payments with digital currencies.

As for applying technological bandages, be careful and tread lightly. As Canadian businessman Craig Bruce has famously said, "Temporary solutions often become permanent problems." The same principle applies here.

In-Store Payments

There are many options for using platforms to handle in-person digital currency payments. Companies can consider adding a digital currency-compatible QR code scanner or Near

Field Communication (NFC) terminal for in-store checkout.* If you use a mobile POS, you may also be able to integrate digital currency payments with your existing system—if the system supports them. Companies like Flexa will integrate with a company's existing payment hardware.

DOING DUE DILIGENCE: EVALUATING THIRD-PARTY PARTNERS, VENDORS, AND SERVICE PROVIDERS

If a company decides to use third-party partners, vendors, and service providers (I'll refer to them collectively as *vendors*), there are a slew of factors to consider in determining which ones are best suited for the organization's needs. It's important to include key organizational constituents in departments such as finance, product design, engineering, and legal as part of the diligence process.

Fraud Prevention

To vet potential vendors, it's best to ask questions such as:

- What tools do they use for fraud detection and prevention?
- In the event of a breach or hack, will they assist the company in protecting its or its customers' assets? How?
- How has the company handled previous fraud incidents? (Remember John Chambers's quote at the start of Chapter 6.)

* NFC lets two devices in close proximity to exchange data and communicate with each other.

Blockchain Analytics

We've already discussed some bad actors in the digital currency space. (Recall "Silk Road" in Chapter 4.) Although the FBI shut down the illegal platform in October 2013, bad actors still use digital currency for money laundering and terrorist financing.

Against this backdrop, it's worth asking if your prospective vendor uses blockchain analytics. These powerful tools analyze and interpret data that's stored on a blockchain. Specialized software and techniques extract information from a blockchain to provide insights into the behavior of users, transactions, and other aspects of a blockchain ecosystem. Chainalysis, CipherTrace, Elliptic, Crystal, and TRM Labs are popular vendors.

Privacy

Questions about privacy include:

- What kind of information is the third-party vendor collecting from your company's customers?
- How will the vendor protect customer information?
- Are these protections sufficient to satisfy the privacy laws in the jurisdictions in which the business operates?

Coins, Costs, and Fees

When vetting potential vendors, consider the following questions:

- Which digital currencies can be used for payments?
- What are the fees for the services?
- How will the company be billed? Typical terms include a flat monthly rate (read: all you can eat) or a per-transaction basis (read: à la carte). Is there flexibility?

Legal and Compliance

Here are a few critical questions to consider:

- Is the vendor registered or licensed in the jurisdictions in which it operates? If not, does it need to be?
- Does the vendor conduct anti-money laundering and sanctions screening?
- What types of internal controls and operational security protocols does the vendor have in place?
- Do the vendor's cybersecurity and encryption protocols conform to generally accepted industry standards?
- Does the vendor use any third parties to audit its services to ensure the safety and soundness of its operations?

Any vendor worth its salt should be able to quickly provide answers to these questions. Ideally, this assurance takes the form of a System and Organization Controls report.*

Customer Experience/Service and Technical Support

A vendor should offer ample customer and technical support. If the business conducts transactions globally, it should ensure that the vendor can provide both types of support across *all* jurisdictions. Say that a firm operates in California and its vendor is based in the Netherlands. The difference in time zones could make real-time support difficult, potentially increasing the time needed to resolve issues that invariably arise. Ditto for language considerations.

And what about customer complaints related to digital currency payments? Will the company handle those itself or rely on

* SOC1 or SOC2 reports focus on outsourced services that may affect how a firm reports its financials.

the vendor? Consider the nature of transactions conducted with digital currencies—and specifically blockchain—discussed at length in Chapter 4. Once a payment goes through, no entity can reverse it. (This is a blockchain feature, not a bug.)

That's not to say, however, that businesses will never allow for the return of goods that were paid for with digital currencies. In these cases, how will the vendor handle such transactions?

Tactical questions to consider include the following:

- What's the technical familiarity of a company's existing and prospective customer base?
- What are the benefits for customers of a company accepting digital currency payments?
- What are the risks (financial or reputational, for example) associated with a company accepting digital currency payments?
- How will the vendor handle failed digital currency transactions?
- How can a company ensure that accepting digital currency payments routinely results in a positive customer experience?
- How will the company address a customer's poor payment experience?

Successfully navigating the world of payments with digital currencies involves much more than just purchasing, configuring, and rejiggering the tech. It's time to talk about money matters.

Finance, Accounting, and Taxes

For the purposes of calculating US federal income tax, the IRS classifies digital currencies as property.[2] Depending on how a company integrates digital currency payment capabilities, it may need to track the value of digital currencies from the time it receives them until it has sold, liquidated, or converted them into fiat currency. (Tax experts call these *taxable events*.) Tracking all these transactions requires a good deal of effort. What's more, given the volatility of certain digital currencies, a company's related transactions may result in unintended tax gains or losses—and it must account for each one of them.

Those that rely upon third-party payment networks need to file additional paperwork. They're required to file Form 1099-K with the IRS and provide a copy to you when the gross payment amount is more than $600.[3]

Payment networks and gateways that fail to capture key information on digital currency-related transactions expose the organization to legal and tax headaches.

In terms of accounting for digital currencies, there are many considerations, the majority of which are beyond the scope of this book. Currently, companies must account for "crypto assets"* as indefinite-lived intangible assets in accordance with ASC 350† (i.e., the assets must be measured at historical cost less impairment) unless the entity is within the scope of the

---

* FASB defines *crypto assets* as digital assets that reside on a distributed ledger based on blockchain or similar technology. This includes digital currencies as defined in this book. FASB is an independent nonprofit organization responsible for establishing accounting and financial reporting standards for companies and nonprofit organizations in the United States, following generally accepted accounting principles (GAAP).

† ASC stands for accounting standards codification. It's a systematic framework used in the US to organize and present accounting standards and principles.

investment-company guidance in ASC 946 or is a certain type of broker-dealer.

In response to criticisms of this traditional intangible asset model,* the Financial Accounting Standards Board (FASB) proposed certain amendments in March 2023. Those amendments seek to improve the accounting for, and disclosure of, crypto assets.

On September 6, 2023, FASB approved a proposed accounting standards update (ASU) related to crypto assets. Under the new guidance, companies must measure crypto assets at fair value, with changes in fair value included in net income in each reporting period. Specifically, companies will be required to:

- Present on the balance sheet the aggregate amount of "crypto assets measured at fair value separately from other intangible assets" that are not measured at fair value.
- Present in net income changes in the fair value of crypto assets separately from changes in the carrying amount (e.g., impairments and amortization) of other intangible assets, including other crypto assets that aren't measured at fair value.
- Classify as cash flows from operating activities those cash receipts from the nearly immediate sale of crypto assets that were "received as noncash consideration in the ordinary course of business (for example, in exchange for the transfer of goods and services to a customer)."[4]

---

* The concerns relate to, among other things, the intangible asset model (1) failing to represent the economics of crypto assets and (2) being an overly complicated method of recognizing impairments by requiring entities to use a crypto asset's lowest observable fair value within a reporting period.

The Issuance of a final ASU is expected during the fourth quarter of 2023. The expected effective date is 2024.

In brief, companies should consult their tax and accounting professionals to understand the potential implications of transacting in digital currencies and holding them on the balance sheet.

The case of MicroStrategy provides an example and some food for thought.

MicroStrategy's Big Digital Currency Play

By way of background, the company made a name for itself in the 1990s by building best-of-breed business-intelligence products. That's not to say, though, that it has remained stagnant since its inception. In fact, its recent transformation is downright stunning.

MicroStrategy's (MSTR) cofounder, former CEO, and current chairman Michael Saylor has been a long-time unabashed digital-currency bull. On his watch, MicroStrategy has purchased at least 140,000 Bitcoin as of this writing.[5]

Against the earlier backdrop of digital currency's unprecedented ascent, Saylor looked like a visionary. The victory laps he took on many of his media appearances only reinforced that notion. Yes, the price of MSTR has fluctuated,[6] but its overall trend has been positive. Saylor even touted Bitcoin as a solution to inflation.[7]

In February 2021, MSTR cracked $1,000 per share—its highest value since the dot-com boom. Not long after, though, Saylor's ambitious gambit gave some investors pause. A few openly questioned his priorities and overall leadership. As the price of Bitcoin plummeted, the cries of Saylor's critics intensified. Specifically, critics questioned whether the company was focusing on its core business of selling software and whether it was wise for MSTR to invest a high percentage of its corporate assets to a volatile digital currency.

Shareholders weren't the only ones expressing concerns. In early 2022, the Securities and Exchange Commission told MSTR "to revise the way it discloses its bitcoin holdings in future filings."[8]

In August of that year, Saylor ceded his CEO role to company president Phong Le. Explaining the move, Saylor cited his desire to "focus more on our bitcoin acquisition strategy and related bitcoin advocacy initiatives."[9]

Fast-forward to May 2023. The company's Bitcoin position had rebounded, allowing it to take a smaller write-off than its accountants and Wall Street had anticipated.[10] That same quarter, MSTR posted a small profit thanks to Bitcoin's 2023 rally.[11]

DIGITAL CURRENCY BALANCE SHEET CONSIDERATIONS

At first, MSTR's investment in Bitcoin may seem irrelevant in the context of reimagining payments. After all, it's not as if the company began accepting digital currency payments from customers. However, that view misses the broader ramifications of the MSTR story.

The company's foray into Bitcoin is fascinating and instructive on a number of levels. Its CEO chose to purchase and hold a significant amount of Bitcoin as a corporate asset, and we need to unpack some of those lessons.

First, holding so much of *any* asset on the books can be risky for any company—digital currencies included. As a general rule, fiat currencies are far more stable than their digital currency counterparts. Don't believe me, though. Figure 10.1 displays the price of Bitcoin over the past dozen years.

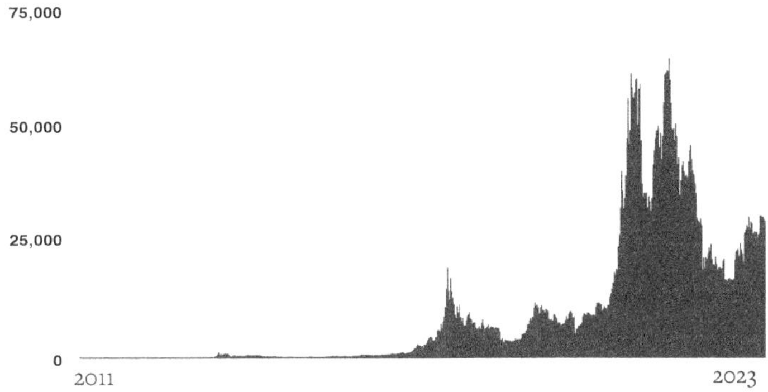

Figure 10.1: Bitcoin Price in USD (June 2011 to June 2023)
Source: Investing.com (https://tinyurl.com/repay-bitcoin)

Saylor's aggressive digital currency strategy may have made some people uncomfortable, but he broached an important issue: fiduciaries of any business routinely seek maximum returns on their assets. To this end, senior leaders should ask themselves the following questions in deciding whether to hold digital currencies on a balance sheet:

- What's the company's risk tolerance?
- If the company accepts digital currencies as payment, does it intend to keep those assets on its books? Or does it plan on immediately converting the digital currency to fiat currency?
- If the company holds the digital currencies on a balance sheet, will it self-custody or use a third-party custodian to store the digital currencies?

- How will digital currencies be accounted for on the company's balance sheet?
- Must the company make disclosures to its shareholders or regulators about digital currency held on a balance sheet?
- How will shareholders and leadership react to rapid and substantial price swings of digital currencies over a short period of time?

This list isn't remotely comprehensive. Think of these questions as starting points. Their answers may change as market conditions warrant and the environment shifts.

CONSUMER CONSIDERATIONS

Let's switch gears. Just like companies, US customers who pay for goods and services with digital currencies will likely owe taxes on those transactions at some point. As of this writing, *spending* digital currencies is effectively tantamount to *selling* it in the eyes of the IRS.

Peter Saddington—whom we met earlier—had to sell his Bitcoin before he could drive away in his dream car. This was a taxable event. Whether John or Jane uses their digital currency proceeds to buy a Lamborghini Huracan or a vanilla latte at Starbucks doesn't matter. In each case, they need to think about Uncle Sam—at least for the time being.

Even in the current hyperpolarized US political environment, I suspect that the status quo will soon change, especially for transactions involving negligible amounts. Bills have been introduced in the US House of Representatives and Senate to create *de minimis* exemptions from capital gains tax for low-value digital currency transactions in day-to-day use.

For now, stay tuned. The situation is fluid.

Other Departments

We're not out of the woods yet. This last section looks at three additional essential functions or departments within a typical organization.

OPERATIONS AND HR

For the sake of simplicity, I'm lumping these two departments together.

Hiring

Depending on the composition of your company, it may need to augment its existing workforce through targeted hires who have familiarity with digital currencies. Areas here may well include engineering, product management, marketing, legal, finance, and compliance:

- **Engineering:** Employees with experience building, improving, and maintaining the applications that process digital currency transactions.
- **Product managers:** Employees who can plan, execute, and track technical projects involving digital currency payments to ensure their successful completion.
- **Business development:** Employees who can generate leads, develop relationships, and create opportunities and business strategies relating to digital currency payments.
- **Legal and compliance:** Employees who can ensure legal and regulatory compliance relating to digital currency payments.

Training

Beyond making strategic hires, organizations may need to provide training to their employees relating to digital currency payments. Areas typically include:

- Education on digital currencies in general and new payments technology.
- How to respond to customers' questions and complaints.
- Compliance obligations for facilitating or accepting digital currency payments.
- Privacy and cybersecurity considerations.
- Accounting and tax considerations.

Employees who lack sufficient knowledge and skills about digital currency payments could expose the company to an array of risks. It's also important to have ongoing training and not adopt a "set-it-and-forget-it" mindset.

MARKETING

As you've seen with the examples of Ralph Lauren and Chipotle, accepting digital currency payments from customers represents a branding opportunity. The potential for differentiation here is significant. Specifically, a company can demonstrate to its customers—and the world at large—its progressive nature. Done properly, it can position the company as forward thinking, tech-savvy, and customer-friendly.

Embracing digital currency payments lets companies tap into existing digital currency communities. These are passionate "groups of individuals with a shared interest in cryptocurrency investing."[12] They usually maintain their own subreddits, Discord servers, X (formerly Twitter), Instagram, and Facebook groups.

Engagement involves creating new environments and communicating with the public as much as possible, making sure that you will be able to address the concerns of the community members.

Questions worth considering include:

- Do our customers want to pay for our products and services with digital currencies?
- Do digital currencies represent an opportunity to support an emerging payment method?
- Do digital currency payments appeal to younger tech-savvy generations?
- Can we gain a competitive advantage?

Parting Thoughts

We've reached the end of our journey. Before putting a bow on this book, a few final words of advice are in order.

First, recognize that the universe of digital currencies and payments is a vast, complicated, and dynamic place. Don't expect to master it in a few weeks or months. Take it from someone who's been ensconced in this world since 2015: there's always more to learn.

Second, expect the unexpected as your organization walks down the aisle. Even late adopters don't always stick the landing. Obstacles abound.

Next, the need for caution around customers and payments is understandable. I get it. Still, you don't want to miss the proverbial boat. Opportunities are plentiful.

Finally, remember the title of this book. We're constantly reimagining payments. It's always been a fluid concept. Corporate and customer needs and preferences change—sometimes overnight. What works now may not work in the next decade, year, or even month. Know as much going in.

If I can help you in any way, please let me know.

Chapter Summary

- Overseeing an organization's digital currency payments initiatives isn't a one-person or one-department job. Planning, prioritization, and monitoring require extensive collaboration between and among departments.
- Making or accepting digital currency payments requires new organizational considerations and new technology—either insourced or outsourced.
- There are finance, tax, accounting, legal, compliance, and information technology implications for digital currency payments. You need to take a holistic approach.

Spreading the Word

"To defend what you've written is a sign that you are alive."

—WILLIAM ZINSSER

Thank you for buying *Reimagining Payments.* I hope that you've enjoyed the pages you just read. Ideally, you've found this book informative, and it has made you think about the opportunities and possibilities for digital currency payments.

If that's true, then perhaps you're willing to help me by doing one or more of the following:

- Writing a book review on Amazon, bn.com, GoodReads, or your blog. The more honest, the better.
- Mentioning this book on Facebook, Reddit, Threads, X (formerly Twitter), Quora, LinkedIn, and other sites you frequent.
- Recommending it to anyone who might find it interesting.
- Giving this book as a gift.

Thanks again.

Michelle Ann Gitlitz
www.michellegitlitz.com
December 1, 2023

Acknowledgments

Writing a book is tougher than I thought and more rewarding than I ever could have imagined. Having an idea and turning it into a book is as hard as it sounds. Thank you, Phil Simon, for helping me bring *Reimagining Payments* to life.

Thank you to my parents, Edward and Janet Gitlitz, whose love, guidance, and unwavering support are always with me.

I'm also grateful to my family and friends for cheering me along during this process: Darlene Gitlitz, Robert Hoffman, Debra Goldsmith, Helene Shapiro, Deysy Maria Santana, Jason Brett, Whitney Kalmbach, Michael Mosier, Claiborne (Clay) Porter, Jane Khodarkovsky, Jennifer B. Lassiter, Hon. J. Christopher Giancarlo, Dana Brakman Reiser, Liesel K. Carvajal, Dave Alderstein, Janeen Sarlin, Dr. Elizabeth Safran, Jess Cheng, Sarah Wetzstein, and all my girlfriends from Cornell University. I'm especially grateful to Noah Lang for his unwavering support.

I couldn't have written this book without the contributions of my current and former colleagues at Flexa Inc.: Trevor Filter, Daniel McCabe, Zachary Kilgore, Tyler Spalding, Sydney Abualy, and the rest of the incredible team. They've supported this book in every way possible, especially Trevor Filter.

The team at Racket Publishing is phenomenal. I'd like to thank Karen Davis, Jessica Angerstein, Vinnie Kinsella, Johnna VanHoose Dinse, and Merlina McGovern.

I owe an enormous debt of gratitude to those in the blockchain, cryptocurrency, and payment ecosystems who have had the vision to test new ideas, develop new ways of thinking, and create innovative technology and applications. I knew nothing about these topics in 2015 when I stumbled upon them. Thank you for embracing and teaching me. I learn something new every day.

About the Author

Michelle Ann Gitlitz, Esq., is a leading expert on cryptocurrencies, digital payments, blockchain, and related technologies. As general counsel of Flexa Inc. and a partner at Arktouros PLLC, she provides strategic legal and advisory services to nascent and established blockchain, cryptocurrency, and financial services companies. She frequently meets with policymakers, legislators, and industry leaders to facilitate intelligent adoption and regulation. Michelle holds degrees from Cornell University and Brooklyn Law School. *Reimagining Payments: The Business Case for Digital Currencies* is her first book. She currently resides in New York City with her two daughters.

Bibliography

Bilton, Nick. *American Kingpin: The Epic Hunt for the Criminal Mastermind Behind the Silk Road*. New York: Penguin, 2018.

Blaug, Mark. *Economic Theory in Retrospect*. Cambridge: Cambridge University Press, 1997.

Chapman, Stanley. *The Rise of Merchant Banking*. London: Routledge, 2013.

Cohen, Adam. *The Perfect Store: Inside eBay*. Boston: Back Bay Books, 2002.

Garten, Jeffrey E. *Three Days at Camp David: How a Secret Meeting in 1971 Transformed the Global Economy*. New York: Harper Collins, 2022.

Giancarlo, J. Christopher. *Cryptodad: The Fight for the Future of Money*. Hoboken, New Jersey: John Wiley & Sons, Inc., 2022.

Goldstein, Jacob. *Money: The True Story of a Made-Up Thing*. New York: Hachette Books, 2020.

Greenberg, Andy. *This Machine Kills Secrets: How Wikileakers, Cypherpunks, and Hacktivists Aim to Free the World's Information*. New York: Dutton, 2012.

Greenberg, Andy. *Tracers in the Dark: The Global Hunt for the Crime Lords of Cryptocurrency*. New York: Doubleday, 2022.

Housel, Morgan. *The Psychology of Money: Timeless Lessons on Wealth, Greed, and Happiness*. Hampshire, Great Britain: Harriman House Publishing, 2021.

Johnson, Steven. *Where Good Ideas Come From: The Natural History of Innovation*. New York: Riverhead Books, 2011.

Leibbrandt, Gottfried, and Teran Natasha De. *The Pay Off: How Changing the Way We Pay Changes Everything*. London: Elliott & Thompson Limited, 2021.

Lewis, Michael. *The Big Short: Inside the Doomsday Machine*. London: Norton & Company, 2011.

Mezrich, Ben. *Bitcoin Billionaires: A True Story of Genius, Betrayal, and Redemption*. New York: Flatiron Books, 2019.

Schwartz, Barry. *The Paradox of Choice: Why More Is Less*. London: Harper Collins, 2004.

Siddiqui, Ahmed. *The Anatomy of the Swipe: Making Money Move*. Potomac, Maryland: New Degree Press, 2020.

Smith, Adam. *The Wealth of Nations*. Blacksburg, Virginia: Thrifty Books, 2009.

Soni, Jimmy. *The Founders: The Story of PayPal and the Entrepreneurs Who Shaped Silicon Valley*. New York: Atlantic Books, 2022.

Sorkin, Andrew Ross. *Too Big to Fail: The Inside Story of How Wall Street and Washington Fought to Save the Financial System—and Themselves*. New York: Penguin, 2010.

Vigna, Paul, and Michael Casey. *The Age of Cryptocurrency: How Bitcoin and the Blockchain Are Challenging the Global Economic Order*. New York: Picador/St. Martin's Press, 2016.

Von Mises, Ludwig. *The Theory of Money and Credit*. Orlando: Signalman Pub, 2009.

Zuboff, Shoshana. *The Age of Surveillance Capitalism: The Fight for a Human Future at the New Frontier of Power*. New York: PublicAffairs, 2019.

Endnotes

Introduction

1 Kara Swisher and Evan Ramstad. "Yahoo! to Announce Acquisition of Broadcast.com for $5.7 Billion." *Wall Street Journal*, April 1, 1999, sec. Tech Center. https://tinyurl.com/2z8e4z9b.

2 Verne Kopytoff. "5 Worst Internet Acquisitions of All Time." *Fortune*. May 21, 2013. https://tinyurl.com/tw658amc.

3 CNN Money. "10 Big Dot.com Flops." March 10, 2010. https://tinyurl.com/2006f4fq.

4 Carvana. "Carvana Leads Industry as Fastest to Sell One Million Vehicles Online Since Founding." Accessed May 17, 2023. https://tinyurl.com/2zkb5nbq.

5 Rashi Maheshwari. "Why Is the Crypto Market Rising Today?" *Forbes*. July 6, 2023. https://tinyurl.com/22rsfj9b.

6 Sean Michael Kerner. "Crypto Winter Explained: Everything You Need to Know." TechTarget. January 26, 2023. https://tinyurl.com/2p274uwe.

Chapter 1

1 Cars Direct. "Why Does a New Car Lose Value After It's Driven Off the Lot?" CarsDirect. March 20, 2011. https://tinyurl.com/y3panmhc.

2 J. H. Cullum Clark. "Why America's Free Market Economy Works Better in Some Places Than Others." George W. Bush Institute. Fall 2019. https://tinyurl.com/2lrtvewk.

3 Charles M. Kahn. "Tokens vs. Accounts: Why the Distinction Still Matters." Federal Reserve Bank of St. Louis. October 5, 2020. https://tinyurl.com/2pfbn4am.

4 Claire Greene, Marcin Hitczenko, Brian Prescott, and Oz Shy. *US Consumers' Use of Personal Checks: Evidence from a Diary Survey*. n.d. https://tinyurl.com/2otbvc9q.

5 American Bankers Association. "1782–1799: A History of America's Banks and the ABA." n.d. https://tinyurl.com/y3ok3fst.
6 Paul Vigna. "Ready to Get Red-Pilled? Money Isn't Real." Information Ratio. May 19, 2023. https://tinyurl.com/2nwtlpc4.
7 Board of Governors of the Federal Reserve System. "The Fed Is US Currency Still Backed by Gold?" Federal Reserve. n.d. https://tinyurl.com/je74llh.
8 Adam Smith. *The Wealth of Nations* (Cosimo Inc., 2007: 35). https://tinyurl.com/bdf4euub.
9 BBC. "The Weimar Republic 1918–1929." Bitesize. 2020. https://tinyurl.com/yzl8qx6t.
10 Michael Bordo. "The Operation and Demise of the Bretton Woods System: 1958 to 1971." CEPR. April 23, 2017. https://tinyurl.com/2jzlm4z8.
11 Sandra Kollen Ghizoni. "Creation of the Bretton Woods System." Federal Reserve Bank of Atlanta. Federal Reserve History. https://tinyurl.com/599u3zr4.
12 "Worth Its Weight in Gold." Idioms Online. May 20, 2021. https://tinyurl.com/2nw7kmt9.
13 US Department of the Treasury. "Legal Tender Status." Internet Archives Wayback Machine. January 26, 2011. https://tinyurl.com/2mavnmoa.
14 Inyoung Hwang. "What Is Fiat Currency? How Is It Different from Crypto?" SoFi. February 16, 2021. https://tinyurl.com/2ej6eajs.
15 Paul Gilkes. "Liberty Dollar's Creator Pens Book Chronicling Its History." CoinWorld. September 25, 2018. https://tinyurl.com/2f38mfrk.
16 Alan Feuer. "Prison May Be the Next Stop on a Gold Currency Journey." *New York Times*. October 25, 2012. https://tinyurl.com/yak7va3z.
17 Brian Summers. "Private Coinage in America." Foundation for Economic Education. July 1, 1976. https://tinyurl.com/2j66ywpl.
18 Barry Dolphin. "Ithaca, NY Launches Its Own Digital Currency to Boost Local Economy." July 22, 2015. https://tinyurl.com/4punzh57.
19 FBI. "Defendant Convicted of Minting His Own Currency." March 11, 2011. https://tinyurl.com/2k2la49c.

20 Tin Fulton Walker & Owen. "United States v. Bernard von NotHaus." Accessed September 26, 2022. https://tinyurl.com/2gdokx7u.

Chapter 2

1 Peri Hartman, Jeffrey P. Bezos, Shel Kaphan, and Joel Spiegel. Method and system for placing a purchase order via a communications network. US Patent 5,960,411A, filed September 12, 1997, and issued September 28, 1999.

2 Peri Hartman, Jeffrey P. Bezos, Shel Kaphan, and Joel Spiegel. Method and system for placing a purchase order via a communications network. US Patent 5,960,411A, filed September 12, 1997, and issued September 28, 1999.

3 U.S. Patent and Trademark Office. "U.S. Patent Statistics Chart Calendar Years 1963–2020." Updated May 2021. https://tinyurl.com/y8832buh.

4 Apple. "Apple Licenses Amazon.com 1-Click Patent and Trademark." Apple Newsroom. September 18, 2000. https://tinyurl.com/myzh6t2.

5 Amazon Just Walk Out. "Learning How the Just Walk Out Technology Experience Works." https://justwalkout.com.

6 Jeffrey Dastin. "Amazon Launches Business Selling Automated Checkout to Retailers." Reuters. March 10, 2020. https://tinyurl.com/wnzfosz.

7 Merriam-Webster. "Payment." Retrieved July 29, 2023. https://tinyurl.com/5n7xudvd.

8 The Law Dictionary. "Pay." https://tinyurl.com/2hdbtzrr.

9 History. "First ATM Opens for Business." This Day in History. n.d. https://tinyurl.com/2j46c8ys.

10 Malcolm Harris. "The War on Cash." Intelligencer. June 22, 2022. https://tinyurl.com/2ja4fp4y.

11 McKinsey & Company. "Next Moves in the Global Payments Ecosystem." October 7, 2022. https://tinyurl.com/2hd7yv72.

12 The Motley Fool. "6 Downsides to Using Cash." The Motley Fool. September 6, 2019. https://tinyurl.com/2odubaca.

13 McKinsey & Company. "Next Moves in the Global Payments Ecosystem." October 7, 2022. https://tinyurl.com/2hd7yv72.

14 Board of Governors of the Federal Reserve System. "Fedwire Funds Services." https://tinyurl.com/yc4strs6.
15 The Federal Reserve. "About the FedNow Service." 2023. https://tinyurl.com/5b2jtca7.
16 Bank for International Settlements. "Committee on Payments and Market Infrastructures." July 2016. www.bis.org/cpmi/publ/d147.pdf.
17 Wim Hordijk. "From Salt to Salary: Linguists Take a Page from Science." NPR. November 8, 2014. https://tinyurl.com/2h795gqu.
18 Frank Holt. "The Invention of the First Coinage in Ancient Lydia." World History Encyclopedia. July 9, 2021. https://tinyurl.com/2z2e7fmx.
19 Hao Zhao, Xiangping Gao, Yuchao Jiang, Yi Lin, Jin Zhu, Sicong Ding, Lijun Deng, and Ji Zhang. "Radiocarbon-Dating an Early Minting Site: The Emergence of Standardised Coinage in China." *Antiquity* 95, no. 383 (August 6, 2021): 1161–78. https://doi.org/10.15184/aqy.2021.94.
20 John Lanchester. "The Invention of Money." *New Yorker*. July 29, 2019. https://tinyurl.com/y695uk5e.
21 GOBankingRates. "Almost Half of Americans Have Not Written a Single Check in the Past Year—Here's What They're Doing Instead." GOBankingRates. January 30, 2023. https://tinyurl.com/2lj75d98.
22 The Federal Reserve Bank of Kansas City. "Banking and Payments Research." May 24, 2023. https://tinyurl.com/2lgcuxzh.
23 Jennifer Collins. "A Short History of the Debit Card." Marketplace. August 18, 2011. https://tinyurl.com/20q8reb4.
24 Diners Club. "History and Legacy." Diners Club International. n.d. https://tinyurl.com/2zdsearx.
25 Eilene Zimmerman. "The Evolution of Fintech." *New York Times*, April 6, 2016, sec. Business. https://tinyurl.com/y6k3as95.
26 World Bank. "Fintech and the Future of Finance." n.d. https://tinyurl.com/2fqa5t9g.

Chapter 3

1 Ohio History Central. "Panic of 1893 Ohio History Central." n.d. https://tinyurl.com/2nd3c3ea.

2 Romer. "Panic of 1893: Estimates of Unemployment During the 1890s." 1984. https://tinyurl.com/2jqf7263.
3 Federal Reserve History. "Banking Panics of the Gilded Age." n.d. https://tinyurl.com/yjwooumb.
4 Benjamin Bromberg. "Temple Banking in Rome." *The Economic History Review*: 128–31. November 1940. https://doi.org/10.1111/j.1468-0289.1940.tb00695.x.
5 Bamber Gascoigne. "History of Banking." December 14, 2009. https://tinyurl.com/2nmyqf6h.
6 Daniel Cohen. "How Hamilton Laid the Foundation for the Federal Reserve." July 2020. https://tinyurl.com/2nzkd59m.
7 U.S. Const. art. 10.
8 David Wheelock. "Overview: The History of the Federal Reserve." Federal Reserve History. September 13, 2021. https://tinyurl.com/20jnyaej.
9 David Wheelock.
10 Craig S. Hakkio. Federal Reserve Bank of Kansas City. "The Great Modernization." November 22, 2013. https://www.federalreservehistory.org/essays/great-moderation.
11 Nikitra Bailey. "Predatory Lending: The New Face of Economic Injustice." American Bar Association. July 1, 2005. https://tinyurl.com/24jh8wrr.
12 Nikitra Bailey.
13 Erin Coghlan, Lisa McCorkell, and Sara Hinkley. "What Really Caused the Great Recession?" Institute for Research on Labor and Employment. September 19, 2018. https://tinyurl.com/yyudnmby.
14 Erin Coghlan, Lisa McCorkell, and Sara Hinkley.
15 NBC News. "New Home Sales Fell by Record Amount in 2007." January 2008. https://tinyurl.com/28kmkx7p.
16 Reuters. "Timeline: A Dozen Key Dates in the Demise of Bear Stearns." March 17, 2008. https://tinyurl.com/yeysvmpe.
17 Corporate Finance Institute. "Lehman Brothers." May 6, 2022. https://tinyurl.com/2m9sd707.
18 Corporate Finance Institute.
19 Amadeo Kimberly. "This Bailout Made Bernanke Angrier Than Anything Else in the Recession." The Balance. November 16, 2020. https://bit.ly/3D8wt12.

20 M. Alex Johnson. "Bush Signs $700 Billion Financial Bailout Bill." NBC News. October 2, 2008. https://tinyurl.com/2fushoft.
21 US Department of the Treasury. "Troubled Assets Relief Program (TARP)." n.d. https://tinyurl.com/ygxsqozc.
22 Christie Les. "Foreclosures Up a Record 81% in 2008." CNN Money. January 15, 2009. https://tinyurl.com/yaj342yq.
23 The Financial Crisis Inquiry Commission. "Final Report of the National Commission on the Causes of the Financial and Economic Crisis in the United States." January 2011. https://www.govinfo.gov/content/pkg/GPO-FCIC/pdf/GPO-FCIC.pdf
24 The Financial Crisis Inquiry Commission.
25 Justin McCarthy. "Americans' Confidence in Banks Still Languishing below 30%." Gallup. June 16, 2016. https://tinyurl.com/2299mfz3.
26 Jeffrey Jones. "Confidence in US Institutions Down; Average at New Low." Gallup. July 5, 2022. https://tinyurl.com/2jbd5pps.
27 Philip Inman. "Wall Street Bonuses Under Fire as Bailed-Out Banks Pay Billions to Executives." *The Guardian*. July 30, 2009. https://tinyurl.com/2b8p7wc7.
28 Julie Stackhouse. "Did the Dodd-Frank Act Make the Financial System Safer?" Federal Reserve Bank of St. Louis. February 17, 2017. https://tinyurl.com/2ake9dme.
29 Jacob Pramuk. "Trump Signs the Biggest Rollback of Bank Rules Since the Financial Crisis." CNBC. May 24, 2018. www.cnbc.com/2018/05/24/trump-signs-bank-bill-rolling-back-some-dodd-frank-regulations.html.
30 Justin Guénette, M. Kose, and Naotaka Sugawara. "Is a Global Recession Imminent? Equitable Growth, Finance, and Institutions Policy Note." September 2022. https://tinyurl.com/2pzfb7p9.
31 Jason Mountford. "2023 Recession Predictions: Is One Coming and How Will It Impact You?" *Forbes*. August 31, 2022. https://tinyurl.com/2930b6hw.
32 Dan Weil. "Economist Roubini: 'Severe' Recession, Financial Crisis Coming." TheStreet. July 25, 2022. https://tinyurl.com/2yf9zr2u.
33 Judy Shelton. "The Not-So-Invisible Hand: Central Banks." *Wall Street Journal*, October 11, 2022, sec. Opinion. https://tinyurl.com/2j9y8vll.

34 Julie Hyman. "Cooling Inflation Data Supportive of a Soft Landing, Economist Explains." Yahoo Finance. July 18, 2023. https://tinyurl.com/2qpodmfd.

35 Anna Wong, Tom Orlik, and Bloomberg. "A Recession Is Still Likely—and Coming Soon." *Fortune*. October 1, 2023. https://tinyurl.com/ysb5pjmu.

Chapter 4

1 Satoshi Nakamoto. "Bitcoin: A Peer-to-Peer Electronic Cash System." 2008. https://bitcoin.org/bitcoin.pdf.

2 Andrea Peterson. "Hal Finney Received the First Bitcoin Transaction. Here's How He Describes It." *Washington Post*. January 3, 2014. https://tinyurl.com/37bxcay5

3 David Chaum. "Blind Signatures for Untraceable Payments." *Advances in Cryptology* 82 (1983): 199–203.

4 Fincash. "What Is eCash?" Retrieved August 7, 2023. https://tinyurl.com/3mvbdwt5.

5 Adam Back. "Hashcash—A Denial of Service Counter-Measure." August 1, 2002. https://tinyurl.com/m7feksx.

6 River Financial. "How Bitcoin Solves the Double Spend Problem." January 27, 2021. https://tinyurl.com/2yx7w98d.

7 Stuart Haber and W. Scott Stornetta. "How to Time-Stamp a Digital Document." *Journal of Cryptology* 3 (1991): 99–111. https://doi.org/10.1007/BF00196791.

8 Marco Iansiti and Karim Lakhani. "The Truth About Blockchain." *Harvard Business Review*. March 6, 2018. https://tinyurl.com/h6n4no6.

9 IBM. "What Are Smart Contracts on Blockchain?" 2022. https://tinyurl.com/ygon9ekz.

10 Ethereum Organization. "Introduction to dApps." July 11, 2023. https://tinyurl.com/5cw9rxyn.

11 Lyle Daly. "How Many Cryptocurrencies Are There?" The Motley Fool. April 19, 2023. https://bit.ly/3TXMPzx.

12 Arjun Kharpal. "Cryptocurrency Luna Now Almost Worthless After Controversial Stablecoin It Is Linked to Loses Peg." CNBC. May 12, 2022. https://tinyurl.com/ycdpmlzf.

13 Todd Spangler. "Sesame Street to Launch First NFTs With VeVe, Starting With Cookie Monster Digital Collectibles at $60 Each." *Variety*. March 13, 2023. https://tinyurl.com/5n87adwe.

14 Heather Lalley. "Three Years In, Chipotle's Lucrative Loyalty Program Continues to Grow." Restaurant Business. May 29, 2022. https://tinyurl.com/275zeuxs.

15 Jessica McLaughlin Stansbury and Alex Hill. "The Blockchain Revolution for Loyalty Programs." OliverWyman. September 19, 2017. https://tinyurl.com/259b9ut9.

16 Michael Bellusci. "Shake Shack Offering Bitcoin Rewards for Customers Using Block's Cash App." CoinDesk. March 4, 2022. https://tinyurl.com/25z2tu8a.

17 Ana Paula Pereira. "Nubank to Launch Loyalty Tokens on the Polygon Blockchain." Cointelegraph. October 19, 2019. https://tinyurl.com/22o3gyea.

18 Sean Michael Kerner. "Colonial Pipeline Hack Explained: Everything You Need to Know." WhatIs.com. April 26, 2022. https://tinyurl.com/28kuaogq.

Chapter 5

1 Federal Trade Commission. "What to Know About Payday and Car Title Loans." Consumer Advice. May 19, 2021. https://tinyurl.com/2knkwayo.

2 Federal Reserve. "The Fed Banking and Credit." Board of Governors of the Federal Reserve System. July 18, 2019. https://tinyurl.com/yd2nrfac.

3 NYC Consumer and Workforce Protection. "Department of Consumer and Worker Protection Research Finds 301,700 NYC Households Are Unbanked." July 9, 2021. https://tinyurl.com/2flbbpmp.

4 FDIC. *2021 FDIC National Survey of Unbanked and Underbanked Households*. July 24, 2023. https://tinyurl.com/y43slkqo.

5 United Nations. "Day of 8 Billion." November 15, 2022. https://tinyurl.com/2d33j4k8.

6 World Bank. "Global Findex Database 2021 Reports Increases in Financial Inclusion Around the World During the COVID-19 Pandemic." July 21, 2022. https://tinyurl.com/2h4yaq26.

7 Global Findex Database. "Global Findex Database 2021 Reports Increases in Financial Inclusion Around the World During the COVID-19 Pandemic." World Bank. July 21, 2022. https://tinyurl.com/32s3fd6x.

8 USAFacts. "Percent of Households That Used an Alternative Financial Service (AFS) in the Past Year." October 23, 2020. https://tinyurl.com/2yoxqmn2.

9 Margaret Coker. "Inside the Controversial Sales Practices of the Nation's Biggest Title Lender." ProPublica. January 19, 2023. https://tinyurl.com/2fekqjco.

10 Dan Murphy. "Economic Impact Payments Uses, Payment Methods, and Costs to Recipients." https://tinyurl.com/2ktpvv7j.

11 Consumer Financial Protection Bureau. "CFPB Research Shows Banks' Deep Dependence on Overdraft Fees." Consumer Financial Protection Bureau. December 1, 2021. https://tinyurl.com/y267al3d.

12 FDIC. *2021 FDIC National Survey of Unbanked and Underbanked Households*. July 24, 2023. https://www.fdic.gov/analysis/household-survey/index.html.

13 The World Bank. "Overview." April 12, 2011. https://tinyurl.com/y88d48vo.

14 Timothy Lyman and Kate Lauer. "What Is Digital Financial Inclusion and Why Does It Matter?" CGAP. March 10, 2015. https://tinyurl.com/2dhslr6z.

15 Marc Andreessen. "Why Software Is Eating the World." Andreessen Horowitz. August 20, 2011. https://tinyurl.com/y2yg7pht.

16 G20. "G20 High-Level Principles for Digital Financial Inclusion." July 23, 2016. https://tinyurl.com/2abrrpcz.

17 Vodafone. "What Is M-Pesa?" 2023. https://tinyurl.com/3uyaswr2.

18 Vodafone.

19 Isaac Mbiti. "The Impact of Mobile Payments on the Success and Growth of Micro-Business: The Case of M-Pesa in Kenya." *Journal of Language, Technology & Entrepreneurship in Africa* 2, no. 1. https://doi.org/10.4314/jolte.v2i1.51998.

20 Tavneet Suri and William Jack. "The Long-Run Poverty and Gender Impacts of Mobile Money." *Science* 354, no. 6317: 1288–92. https://doi.org/10.1126/science.aah5309.

21 Ukraine's Official Twitter Account. Twitter. February 26, 2022. https://tinyurl.com/yam8asv9.

22 Romain Dillet. "How Ukraine Is Using Crypto Donations." TechCrunch. February 28, 2022. https://tinyurl.com/ya5bc7tc.

23 Etherscan. "Ukraine Crypto Donation." Ethereum (ETH) Blockchain Explorer. February 26, 2022. https://tinyurl.com/2g2x4bmb.

24 Romain Dillet. "How Ukraine Is Spending Crypto Donations." TechCrunch. March 2, 2022. https://tinyurl.com/ybz32nvu.

25 Amitoj Singh. "Ukraine Bought Weapons, Drones with Crypto Donations." CoinDesk. August 17, 2022. https://tinyurl.com/2louqbvv.

26 Emily Rome. "How Bitcoin Can Help Bridge Afghanistan's Gender Gap." Inverse. June 27, 2019. https://tinyurl.com/2lfe307e.

27 Ariana News. "WFP Survey Finds About 98% of Afghans Not Getting Enough Food." December 15, 2021. https://tinyurl.com/2fsgku2j.

28 Ali Latifi, Tamana Ayazi, and Mujtaba Haris. "Facing Hunger, Some Desperate Afghans Are Selling Their Kidneys for Money." Business Insider. January 20, 2022. https://tinyurl.com/2p3nh45l.

29 Emanuela-Chiara Gillard. "Humanitarian Exceptions: A Turning Point in UN Sanctions." Chatham House—International Affairs Think Tank. December 20, 2022. https://tinyurl.com/2lnuhagd.

30 Joshua Zitser. "Impoverished Afghan Women Are Receiving Emergency Aid in Crypto as the Taliban Limits Cash Withdrawals and Millions Go Hungry." Business Insider. January 23, 2022. https://tinyurl.com/2luwthzf.

31 Lee Fang. "Starving Afghans Use Crypto to Sidestep US Sanctions, Failing Banks, and the Taliban." The Intercept. January 19, 2022. https://tinyurl.com/yax8a4mg.

Chapter 6

1 Irini Miyashiro. "Equifax Data Breach." Seven Pillars Institute. April 30, 2021. https://tinyurl.com/yd5eh8nm.

2 Josh Fruhlinger. "Equifax Data Breach FAQ: What Happened, Who Was Affected, What Was the Impact?" CSO Online. February 12, 2020. https://tinyurl.com/v9x2dq5.

3 BBC News. "Yahoo 2013 Data Breach Hit 'All Three Billion Accounts.'" BBC News, October 3, 2017, sec. Business. https://tinyurl.com/2hwflfxu.

4 Chris Arnold. "Equifax CEO Richard Smith Resigns After Backlash Over Massive Data Breach." NPR. September 26, 2017. https://tinyurl.com/2hje852y.

5 Federal Trade Commission. "Equifax Data Breach Settlement." July 11, 2019. https://tinyurl.com/2foogyxm.

6 Equifax. "Equifax Data Breach Settlement." August 8, 2018. https://tinyurl.com/y5dyn6kl.

7 Farhad Manjoo. "Seriously, Equifax? This Is a Breach No One Should Get Away With." *New York Times*, September 8, 2017, sec. Technology. https://tinyurl.com/y7mjcp6k.

8 Tracy Kitten. "Chase Breach: 465,000 Accounts Exposed." Bank Info Security. December 5, 2013. https://tinyurl.com/2khnr227.

9 Renata Geraldo. "No Prison for Seattle Hacker Behind Capital One $250M Data Breach." *Seattle Times*. October 4, 2022. https://tinyurl.com/2jdw7ur2.

10 ITRC. "First American Financial Breach Exposes Millions of Complete Identities." ITRC. May 5, 2022. https://tinyurl.com/2mtukjfr.

11 David E. Sanger. "Marriott Concedes 5 Million Passport Numbers Lost to Hackers Were Not Encrypted." *New York Times*. January 4, 2019. https://tinyurl.com/y8k52uu2.

12 Chelsey Cox. "US Banks Processed Roughly $1.2 Billion in Ransomware Payments in 2021, According to Federal Report." CNBC. November 21, 2022. https://tinyurl.com/2dqbb30e.

13 US Department of Labor. "Guidance on the Protection of Personal Identifiable Information." January 21, 2016. https://tinyurl.com/y6vc4ay3.

14 Facebook. "Leading Websites Offer Facebook Beacon for Social Distribution." Internet Archive Wayback Machine. February 14, 2008. https://tinyurl.com/2zgnhuga.

15 Betsy Schiffman. "Facebook CEO Apologizes, Lets Users Turn Off Beacon." *Wired*. December 5, 2007. https://tinyurl.com/2g852bwo.

16 Phil Simon. "Big Data Lessons from Netflix." *Wired*. March 15, 2014. https://tinyurl.com/lvp3uwy.

17 Dan Fletcher. "How Facebook Is Redefining Privacy." *Time*. Accessed May 20, 2010. https://tinyurl.com/2kj66kfc.

18 Estelle Laziuk. "Daily IOS 14.5 Opt-in Rate." Flurry. April 29, 2021. https://tinyurl.com/ygmht74d.

19 Ankita Garg. "Facebook Says Revenue to Be Down by $10 Billion Due to Apple Privacy Changes." *India Today*. February 3, 2022. https://tinyurl.com/ycn5mrlo.
20 Andrew Perrin. "Half of Americans Have Decided Not to Use a Product or Service Because of Privacy Concerns." Pew Research Center. April 14, 2020. https://tinyurl.com/y4fhqt94.
21 Brooke Auxier, Lee Rainie, Monica Anderson, Andrew Perrin, Madhu Kumar, and Erica Turner. "Americans and Privacy: Concerned, Confused and Feeling Lack of Control Over Their Personal Information." Pew Research Center. November 15, 2019. https://tinyurl.com/v9lvl5x.
22 IMF. "Anti-Money Laundering/Combating the Financing of Terrorism (AML/CFT) Topics." 2019. https://tinyurl.com/2p2b9elg.
23 McKinsey & Company. "New Trends in US Consumer Digital Payments." October 26, 2021. https://tinyurl.com/y8t333vp.
24 Merriam-Webster. "Custody." n.d. https://www.merriam-webster.com/dictionary/custody.
25 Crypto.com. "What Is a Crypto Wallet? A Beginner's Guide." April 26, 2022. https://tinyurl.com/280225yk.
26 Alisha Chhangani. "Snapshot: Which Countries Have Made the Most Progress on CBDCs So Far in 2023." Atlantic Council. September 1, 2023. https://tinyurl.com/yjvf6x5b.
27 Federal Reserve Bank of Boston. "Project Hamilton Phase 1 Executive Summary." February 3, 2022. https://tinyurl.com/ybhqlps6.
28 Trishala Chokhani. "A Deep Dive into the Shrinking Attention Span." ELearning Industry. February 23, 2023. https://tinyurl.com/2hugzh2e.
29 Binance Academy. "Finality." October 25, 2020. https://tinyurl.com/2lkqh4uc.
30 Visa Fact Sheet. https://tinyurl.com/ymrhps3n.
31 Netflix. "Chaos Monkey." December 28, 2017. https://tinyurl.com/y2p8zxk8.
32 Michael S. Barr. "The Federal Reserve's Role in Supporting Responsible Invocation." Board of Governors of the Federal Reserve System. September 8, 2023. https://tinyurl.com/vme478hz.
33 Sarah Meiklejohn, Marjori Pomarole, Grant Jordan, Kirill Levchenko, Damon McCoy, Geoffrey M. Voelker, and Stefan Savage. "A

Fistful of Bitcoins: Characterizing Payments Among Men with No Names." (23 October 2013). Proceedings of the 2013 Conference on Internet Measurement Conference. IMC '13: 127–140. doi:10.1145/2504730.2504747.

34 Robert McMillan. "The U.S. Cracked a $3.4 Billion Crypto Heist—and Bitcoin's Anonymity." *Wall Street Journal*. April 12, 2023. https://tinyurl.com/2kjtgr2y.

35 Daniel Gorfine and Michael Mosier. "Opinion: Stablecoin and Other Digital Assets Are Falsely Framed as a Choice Between Personal Privacy and National Security. We Can Have Both." Market Watch. July 23, 2022. https://tinyurl.com/bdz2nu5j.

Chapter 7

1 MyWallSt Staff. "A Brief History of Apple's Payment Revolution." The Motley Fool. September 3, 2019. https://tinyurl.com/2k3zbwex.

2 Cadie Thompson. "Can Apple Win the Mobile Wallet War?" CNBC. November 3, 2014. https://tinyurl.com/2h264vcq.

3 Jonny Evans. "Driven by Pandemic, the US Enters the World of Apple Pay." *Computerworld*. February 25, 2021. https://tinyurl.com/yatecj3t.

4 Pew Research. "More Americans Are Joining the 'Cashless' Economy." Pew Research Center. October 5, 2022. https://tinyurl.com/repay-pew.

5 "2022 Accessible Version of Trends in Noncash Payments." Board of Governors of the Federal Reserve System. 2022. https://tinyurl.com/2xmb39lg.

6 Robert M. Hunt. "A Century of Consumer Credit Reporting in America." FRB Philadelphia Working Paper No. 05-13. SSRN. June 1, 2005. https://tinyurl.com/ybgb7q8z.

7 Thomas A Darkin. "Credit Cards: Use and Consumer Attitudes, 1970–2000." https://tinyurl.com/2z4dejak.

8 Bob Musinski. "The History of Credit Cards." *US News & World Report*. January 26, 2021. https://tinyurl.com/2lksluvb.

9 US Government Accountability Office. "Credit Cards: Pandemic Assistance Likely Helped Reduce Balances, and Credit Terms Varied Among Demographic Groups." September 29, 2023. https://tinyurl.com/5249p4vs.

10 Clearly Payments. "The Growth of the Credit Card Industry in 2023." https://tinyurl.com/ms39kkve.
11 BigCommerce and Google. "Global Consumer Report: Current and Future Shopping Trends." 2022. https://tinyurl.com/2ypqano5.
12 Chipotle. "Burritos or Bitcoin: Chipotle to Give Away $200k in Free Burritos and Bitcoin to Celebrate National Burrito Day." March 21, 2021. https://tinyurl.com/2jka6vxc.
13 Marty Swant. "How Chipotle Is Using Crypto and Gaming as Gateways to Loyalty." Digiday. July 27, 2022. https://tinyurl.com/20un4ofa.
14 Chris Kelly. "Chipotle Lets Roblox Players Roll Virtual Burritos into Real Food." Marketing Dive. April 5, 2022. https://tinyurl.com/2frm9mdm.
15 Matthew Denis. "Chipotle's Metaverse Burrito Builder: Is It Worth the Time?" The Manual. May 17, 2022. https://tinyurl.com/20qydoy9.
16 Chipotle Mexican Grill. "Chipotle Encourages Fans to 'Buy the Dip' with New $200,000+ Crypto Game and 1-Cent Guac for National Avocado Day." Cision PR Newswire. July 25, 2022. https://tinyurl.com/22nvg7sj.
17 Elizabeth Napolitano. "Chipotle Now Accepting Cryptocurrency Payments at US Locations." CoinDesk. June 2, 2022. https://tinyurl.com/2qrhatto.
18 Marty Swant. "How Chipotle Is Using Crypto and Gaming as Gateways to Loyalty." Digiday. July 27, 2022. https://tinyurl.com/20un4ofa.
19 Beth Costa and Neeko Gardner. "How Gen Z Shops and Pays: A Guide for Merchants." https://tinyurl.com/2lqdtze9.
20 Sheena S. Iyengar and Mark R. Lepper. "When Choice Is Demotivating: Can One Desire Too Much of a Good Thing?" *Journal of Personality and Social Psychology* 79, no. 6 (2000): 995–1006. https://tinyurl.com/2m3f2vqv.
21 Scott Galloway. "Consumers don't want more choice, they want to be more confident in the choices presented." Twitter. May 30, 2022. https://tinyurl.com/2ko6er6y.

22 Steff Yotka. “Balenciaga and Fortnite Team Up for a Digital-to-Physical Partnership.” September 20, 2021. http://tinyurl.com/bdvw3a84.

23 Brian Quarmby. “Walmart CTO Says Crypto Will Become a ‘Major’ Payments Disruptor.” Cointelegraph. October 18, 2022. https://tinyurl.com/2qptmd4n.

24 Condé Nast. “Ralph Lauren’s New Web3-Centric Miami Store Accepts Crypto.” Vogue Business. April 4, 2023. https://tinyurl.com/2yffsouy.

Chapter 8

1 Sebastian Rupley. “Meet the Buyer of the Broken Laser Pointer.” eBay. September 11, 2015. https://tinyurl.com/23smxawf.

2 Brian Bergstein. “eBay Buys PayPal in $1.3B Stock Deal.” Associated Press. July 8, 2002. https://tinyurl.com/2qtzndgl.

3 Brian O’Connell. “History of PayPal: Timeline and Facts.” TheStreet. January 2, 2020. https://tinyurl.com/2zszylgc.

4 Jimmy Soni. *The Founders: The Story of PayPal and the Entrepreneurs Who Shaped Silicon Valley*. Atlantic Books. 2022.

5 Failory. “What Was Amazon Auction and Why Was It Discontinued?” April 10, 2021. https://tinyurl.com/2ggwfe4h.

6 Brian Bergstein. “eBay Buys PayPal in $1.3B Stock Deal.” Associated Press. July 8, 2002. https://tinyurl.com/2qtzndgl.

7 Swift. “History.” February 15, 2016. https://tinyurl.com/27mhf7om.

8 Fifth Third Bancorp. “Ninety-Six Percent of Americans Are So Impatient They Knowingly Consume Hot Food or Beverages That Burn Their Mouths, Finds Fifth Third Bank Survey.” Cision PR Newswire. January 27, 2015. https://tinyurl.com/yea7nlqf.

9 Martin Walker. “Do Six Percent of Financial Transactions Sent via the Swift System Really Fail?” *LSE Business Review*. November 4, 2019. https://tinyurl.com/2abqa2j4.

10 Swift. “Swift Unlocks Potential of Tokenisation with Successful Blockchain Experiments.” August 31, 2023. http://tinyurl.com/492ke7zn.

11 BNY Mellon. *Institutional Investing 2.0: Migration to Digital Assets Accelerates*. https://tinyurl.com/bny-assets.

12 Cassie Bottorff and Kimberlee Leonard. "Credit Card Processing Fees (2022 Guide)." Forbes Advisor. March 24, 2023. https://tinyurl.com/2r3h4uwy.

13 Jessica Merritt. "What Are Interchange Rates? [And 10 Ways to Reduce Them]." Upgraded Points. July 24, 2023. https://tinyurl.com/yc7vrrtj.

14 Lyle Daily. "Average Credit Card Processing Fees and Costs in 2023." The Ascent. September 1, 2023. https://tinyurl.com/3eesdh2r.

15 Natasha Gabrielle. "It's Illegal to Charge Credit Card Processing Fees in These 5 States. Here's What to Do." *USA TODAY*. April 5, 2021. https://tinyurl.com/2k7wj4yv.

16 Nilson Report. "Merchant Processing Fees in the United States—2021." March 31, 2022. https://tinyurl.com/2dvfdk9d.

17 FDIC. *How America Banks: Household Use of Banking and Financial Services 2019 FDIC Survey*. https://tinyurl.com/yepbqg33.

18 BIS. "Committee on Payments and Market Infrastructures Interlinking Payment Systems and the Role of Application Programming Interfaces: A Framework for Cross-Border Payments Report to the G20." https://tinyurl.com/2n7fovex.

19 Kristalina Georgieva. "Confronting Fragmentation: How to Modernize the International Payment System." International Monetary Fund. May 10, 2022. https://tinyurl.com/2e6eogjl.

20 World Bank. "Remittances—A Gateway to Sustainable Development." December 20, 2022. https://tinyurl.com/2q5eh3nq.

21 Allied Market Research. "B2B Payments Market Research 2031." November 22, 2020. https://tinyurl.com/2yu33pd7.

22 McKinsey & Company. "Global Banking Practice: The 2020 McKinsey Global Payments Report." https://tinyurl.com/29zr8ndg.

23 Bret Miller. "Digital Disruption in Payments." Lazard Asset Management. April 15, 2019. https://tinyurl.com/2o3ulzzw.

24 Bloomberg. "TassatPay Exceeds $1 Trillion in Real-Time Transactions." February 8, 2023. http://tinyurl.com/4sv96maj.

25 Joseph Poon and Thaddeus Dryja. "The Bitcoin Lightning Network: Scalable Off-Chain Instant Payments." https://tinyurl.com/q54gnb4.

26 Digital Currency Initiative. "Layer 2 | The Lightning Network." https://tinyurl.com/2m3osrvb.

27 Teresa Xie. "Bitcoin's Lightning Network Scaling Solution Seeks Resurgence After Losing Way." October 16, 2023. https://tinyurl.com/25zrae7m.

28 Alyssa Hertig. "David Marcus's Lightspark Unveils Bitcoin Lightning Platform for Business." Decrypt. April 11, 2023. https://tinyurl.com/25av7sej.

29 Tomio Geron. "How the Lightning Network Could Speed Up Bitcoin Payments Protocol." Protocol. January 31, 2022. https://tinyurl.com/2noep96d.

30 Jamie McGeever, Marcela Ayres, and Carolina Mandl. "Brazil Launches 'Pix' Instant Payments System, Whatsapp to Enter 'Soon.'" Reuters. November 16, 2020. https://tinyurl.com/2hcbxbq5.

31 Jamie McGeever, Marcela Ayres, and Carolina Mandl. "Brazil Launches 'Pix' Instant Payments System, Whatsapp to Enter 'Soon.'" Reuters, November 16, 2020, sec. Technology News. https://tinyurl.com/2hcbxbq5.

32 Maria Eloisa Capurro and Shannon Sims. "Brazil's Central Bank Built a Mobile Payment System with 110 Million Users." Bloomberg. October 6, 2021. https://tinyurl.com/2f6hbyxx.

33 UTORG. "New Payment Method: PIX." September 22, 2022. https://tinyurl.com/2m3xeqd4.

34 TBR Newsroom. "Brazil Breaks a New Record for Daily Instant Transactions." The Brazilian Report. October 9, 2023. https://tinyurl.com/2t4866rr.

35 Fabio Plein and Nana Murugesan. "Introducing Exciting New Updates for Our Brazilian Community." March 21, 2023. https://tinyurl.com/2025wvh3.

36 Ripple. "Cross-Border Payments: Settlements in Seconds, Not Days." https://tinyurl.com/ppth7udk.

37 David B. Yoffie and George Gonzalez. "Ripple: The Business of Crypto." HBR Store. February 18, 2020. https://tinyurl.com/2yj703yr.

38 Global Financial Markets Institute. "The Oxygen of the Financial Markets: Liquidity Planning During COVID-19." October 30, 2020. https://tinyurl.com/26ledh29.

39 BDC. "What Is Liquidity?" September 27, 2021. https://tinyurl.com/283sux76.

40 David B. Yoffie and George Gonzalez. "Ripple: The Business of Crypto." HBR Store. February 18, 2020. https://tinyurl.com/2yj703yr.

41 Project mBridge. "Multiple CBDC (MCBDC) Bridge." September 2022. http://tinyurl.com/2eq54ckt.

42 Derek Andersen. "BIS Releases Full Report on mBridge Wholesale CBDC Platform After Successful Pilot." Cointelegraph. October 26, 2022. https://tinyurl.com/yc42fwme.

Chapter 9

1 House Financial Services Committee. "About Us." May 14, 2011. https://tinyurl.com/s7um82d.

2 Global Data. "ShieldSquare Captcha." November 15, 2022. https://tinyurl.com/2gc65ygz.

3 Polaris Market Research. "Cryptocurrency Market Size Global Report, 2022–2030." May 15, 2023. https://tinyurl.com/2g7h20nh.

4 Tanaya Macheel. "Goldman Sachs Launches In-House Incubator." Tearsheet. March 15, 2018. https://tinyurl.com/2m86yprh.

5 PayPal. "PayPal Launches New Service Enabling Users to Buy, Hold and Sell Cryptocurrency." PayPal Newsroom. October 21, 2020. https://tinyurl.com/y3cpcxtn.

6 PayPal. "PayPal Launches U.S. Dollar Stablecoin." August 7, 2023. https://tinyurl.com/cpft4e8r.

7 Coinbase. "The State of Crypto: Corporate Adoption." Second Quarter 2023. https://tinyurl.com/3xzed7m4.

8 Ray Ndlovu. "Zimbabwe Banks May Offer Loans Backed by New Digital Money." Bloomberg. May 18, 2023. https://tinyurl.com/2qm7adm6.

9 Elizabeth Howcroft. "Deutsche Bank to Hold Crypto for Institutional Clients." Reuters. September 14, 2023. https://tinyurl.com/mr29s5cj.

10 Krisztian Sandor and Ekin Genç. "The Fall of Terra: A Timeline of the Meteoric Rise and Crash of UST and LUNA." May 20, 2022. https://tinyurl.com/yc56euhz.

11 Coinbase. "Terra (LUNA) Price, Charts, and News." Accessed July 4, 2022. https://www.coinbase.com/price/terra-luna.

12 Paul Kiernan. "Yellen Renews Call for Stablecoin Regulation After TerraUSD Stumble." *Wall Street Journal.* May 10, 2022. https://tinyurl.com/yv57x24f.

13 Paige Rooney and Kate Tortorelli. "Sam Bankman-Fried's Alameda Quietly Used FTX Customer Funds Without Raising Alarm Bells, Say Sources." CNBC. November 13, 2022. https://tinyurl.com/2bu5fk4h.

14 The Block Research. Financial Services Committee. Congress.gov; Senate.gov.

15 Scribd. "The State of Crypto: Corporate Adoption." Second Quarter 2023. https://tinyurl.com/3xzed7m4.

16 European Securities and Markets Authority. "Markets in Crypto-Assets Regulation (MiCA)." June 15, 2023. https://tinyurl.com/2l7n9mg8.

17 HM Treasury. "Government Sets Out Plan to Make UK a Global Cryptoasset Technology Hub." April 4, 2023. https://tinyurl.com/mr35hddk.

18 Camomile Shumba. "UK Crypto, Stablecoin Laws Approved by Parliament's Upper House." CoinDesk. June 19, 2023. https://tinyurl.com/2dyh536t.

19 Sandali Handagama. "Dubai Mandates Licensing for Crypto Companies as It Sets Out Regulatory Requirements." February 7, 2023. https://tinyurl.com/3f4c2knx.

20 Miles Brooks. "Colorado Is Accepting Crypto for Tax Payments—It Could Be a Mess or a Shining Example." Cointelegraph. October 6, 2022. https://tinyurl.com/2mrfr8pz.

21 Lucas Manfredi. "Equinox's NYC Fitness Clubs to Accept Crypto Payments." FOXBusiness. May 3, 2022. https://tinyurl.com/2eesmdey.

22 Flexa. "Sheetz Becomes First Convenience Store Chain to Accept Bitcoin." PR Newswire. May 27, 2021. https://tinyurl.com/2onnlqwl.

23 Finloop. "Accepting Cryptocurrency on Shopify." July 28, 2022. https://tinyurl.com/mryufa6e.

24 Daniel Ruby. "Shopify Statistics 2022: Revenue, Facts & Trends." Demandsage. November 7, 2022. https://tinyurl.com/2fe5vf5l.

25 Helene Braun. "Shopify Customers Can Now Pay in USDC via Solana Pay." August 23, 2023. https://tinyurl.com/4u9cvh37.

Chapter 10

1 Stripe. "Payment Methods API." July 1, 2019. https://tinyurl.com/yx9zun8h.
2 Douglas Fahey. "Notice 2014-21." IRS. https://tinyurl.com/y43rt8bn.
3 IRS. "Understanding Your Form 1099-K." Internal Revenue Service. December 25, 2019. https://tinyurl.com/2ymn9r6z.
4 FASB. "Proposed Accounting Standards Update—Intangibles—Goodwill and Other—Crypto Assets (Subtopic 350-60): Accounting for and Disclosure of Crypto Asset." https://tinyurl.com/4mb5kxr2.
5 Jamie Redman. "MicroStrategy's Bitcoin Holdings Reach 140,000 BTC After Acquiring 1,045 More Bitcoins." Bitcoin.com. April 5, 2023. https://tinyurl.com/2xvcd6z2.
6 Yahoo! Finance. "MicroStrategy Incorporated (MSTR) Stock Price, News, Quote & History." Accessed June 11, 2023. https://tinyurl.com/24fwnlkn.
7 Jordan Major. "Here's Why Bitcoin Is a Solution to Inflation, According to Michael Saylor." Finbold. April 20, 2022. https://tinyurl.com/28u7b3uh.
8 Mark Maurer. "MicroStrategy to Continue Buying Bitcoin Despite Market Tumble, CFO Says." *Wall Street Journal*. January 25, 2022. https://tinyurl.com/23zw56wa.
9 MacKenzie Sigalos. "MicroStrategy CEO Saylor Moves to Chairman Role, Focusing on Strategy and Bitcoin." CNBC. August 2, 2022. https://tinyurl.com/2argaws3.
10 Aoyon Ashraf. "Michael Saylor's MicroStrategy Books a Much Smaller Bitcoin Impairment Charge." CoinDesk. May 1, 2023. https://tinyurl.com/252omuym.
11 Olga Kharif. "MicroStrategy Posts Profit on Benefit Tied to Bitcoin Holdings." Bloomberg. May 1, 2023. https://tinyurl.com/2627yobn.
12 Geri Mileva. "Top 30 Crypto Communities to Join Right Now." Influencer Marketing Hub. June 12, 2022. https://tinyurl.com/25dqestn.

Index

E

F

www.ingramcontent.com/pod-product-compliance
Lightning Source LLC
Chambersburg PA
CBHW030458210326
41597CB00013B/717

* 9 7 9 8 9 8 7 8 6 4 9 1 3 *